Blokes and Sheds

MAKERS, BREAKERS & FIXERS

MARK THOMSON

HarperCollins*Publishers*

HarperCollins*Publishers*

First published in Australia in 2007
by HarperCollins*Publishers* Australia Pty Limited
ABN 36 009 913 517
www.harpercollins.com.au

Copyright © Mark Thomson 2007

The right of Mark Thomson to be identified as the author and photographer of this work has been asserted by him in accordance with the *Copyright Amendment (Moral Rights) Act 2000*.

HarperCollins*Publishers*
25 Ryde Road, Pymble, Sydney, NSW 2073, Australia
31 View Road, Glenfield, Auckland 10, New Zealand

National Library of Australia Cataloguing-in-Publication data:

Thomson, Mark, 1955– .
 Makers, breakers and fixers.
 ISBN 978 07322 8343 8 (pbk.).
 1. Toolsheds – Social aspects – Australia – Anecdotes. 2. Men – Australia – Social life and customs – Anecdotes.
 3. Australia – Social life and customs – Anecdotes. I. Title.
392.36

Cover photograph by Mark Thomson of Terry and his shed
Cover and internal design by Darren Holt
Printed and bound in Australia by Griffin Press on 130gsm Matt Art

6 5 4 3 2 1   07 08 09 10 11

The answer's in our own backyard.

But what's the question? This book is dedicated to all those makers, breakers and fixers with the curiosity and creativity to ask.

CONTENTS

Chapter 3
THE CORRUGATED BROTHERHOOD

Chapter 4
THE CAR SHED

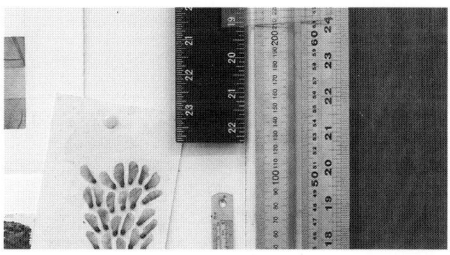

Chapter 5
THE CREATIVE SHED

Chapter 6
ON THE TOOLS

Chapter 7
THE SHED AS A PALACE

ACKNOWLEDGMENTS

The Honourable Stephanie Key MP, Bob Sneath MLC, Ivan Venning MP, Dave Whyte of Gizmo Tools, Chris Barrett of *Queensland Country Life* newspaper, Linda Nathan of *Australian Wood Review*, Jonathan Leahey, Lloyd Linson-Smith, Greg Mackay, John Hardy (Qld DPI), Chris Forsberg, Pat Shears, Ray Sammut of Japoonvale, Pat Guilfoyle, Ms Christine Haines, Paul Wildman (the bricoleur), George Lewin, David Archer, Des and Elaine Nitschke, Kim Hebenstreit, Brian Tanti (a true gent), the members of the Hand Tool Preservation Association of Australia, the Harfull family of Mil Lel, Helen Frances (NZ), Mario Raso, John Taylor, Geoff Buckby, Chris Knight, Barrie Mawby, Joe Garreffa, Mark Lomman, Devon Amber, Peter and Jackie Bandy, Ralph Goss (Institute of Automotive Mechanical Engineers), Reg Ingold, Ed Wilson, Russell Moor (Orange), Nic Marinos, Bob Nelson of Mary McKillop Outreach, Peter Wiley of Millthorpe Museum, Avia Colton, Karen Genoff, Peter Tucker of Renmark, *The Renmark Pioneer* newspaper, Pearl Pannikar, Richard Harries (Old Errowanbang), Alistair Traill, Chris Rule, *ReNew* magazine, Mr Laurie Leibhardt, Geoff Barnes, Frank Bauer, Neil Weidenbach, Tony Kearney, Andrew King of The Bull's Run, Mike Petersen, Glen Moon, Joe Mizzi of Ingham (Qld) and the many shed brothers and others I visited and spoke to in the course of putting together this book.

Special mention must be made of Ian and Noelle Tolley of Renmark (SA), Mr Kevin Gilders of Notting Hill (Vic), Mr Tony Kajper of Coober Pedy, Mr Richard Fewster (Porter Novelli), the great Ron Edwards of Kuranda (Qld), The Williamstown Shed Club and Mike Hamilton in particular, Mr Richard Hunt and also Kent Mayo of McCrossin's Mill, Uralla (NSW).

Mr John Bennett, mechanical genius, of Melbourne, who always challenged me to overhaul the definitions of this project.

Amruta Slee and Jennifer Blau of HarperCollins for their patience.

Richard Hudson, ABC FNQ, Bruce Reynolds, ABC Central West NSW, Ms Deb Tribe, ABC Riverland and Phillip Adams for his interest.

Robi Watt deserves thunderous applause for her dedication to the SBS Blokes and Sheds project, aided admirably by Sara Wishart, Aaron Gully and Jennifer Crone.

The Usual Suspects who made travelling to other cities and places bearable — Graeme Isaac, Winsome Bernard, Kim Batterham, Michael Snelling, Suhanya Raffel, Merridy Isaac and the girls, Keith Noble, Tania Dennis and Robina Hobbs.

Thank you all for your help, open minds and immense kindness.

Introduction
THE NATIONAL SECRET

When, about a decade ago, I produced the book *Blokes & Sheds*, it hit a nerve and sold a lot of copies. That backyard success was very pleasing but also a little troubling. It created the impression that I actually knew some sort of secret about men, or corrugated iron, or national identity. All bullshit, of course.

In the intervening period I have had time to give this suspected secret some thought. I've seen a lot more sheds, maybe too many for my own good. I've kicked a lot of tyres and drawn a few mudmaps. I've chewed the fat with shed scientists, corrugationalists and every other shade of the true shed brotherhood. We ask each other: What is a good shed all about? If there is a secret, we all know it, but hardly ever put it into words.

To save time I'll tell you right now what I've discovered. It comes down to a formula:

SHED = PRACTICAL = PURPOSE = MEANING

Shed = practical, which = purpose, which = meaning.

It's as simple as that: the meaning of life in six words.

If that's all you wanted to know, you can put this book down and go off and do something useful. Everything else in the book stems from this formula or explains it a bit further.

Another way of explaining what I've discovered would be to start from the idea that every building, like it or not, conveys a set of values. The courthouse or parliament house aims to impress on us the power of the state. The cathedral signals spiritual power, university buildings hope to symbolise the richness of human knowledge, and corporate headquarters want to suggest financial stability.

So what values does a shed convey?

A shed is a mortal, often flimsy building. From the moment of its construction, its days on this earth are numbered. It is not elevated or weighed down by the need to represent grand and unchanging values. In fact it may enjoy being slightly disreputable — or very disreputable indeed.

The main value a shed represents is the importance of usefulness. A shed is a building stripped down to its purpose. It can take any shape, be in any condition, even look likely to blow away in the next big storm, as long as it does what it's supposed to do.

Some sheds are loaded with extra significance as well, whether the giant 19th-century shearing shed which we have embraced as a national symbol, or a modest backyard shed built two years ago that's full of personal meaning to its owner.

In some sheds, the personality of their owners seems to hover in the very air, or reside in the grain of the workbench along with the linseed oil, paint stains and chisel marks.

The essential quality of a good shed is impossible to pin down: you just know it. When you go in there you feel a sort of inner sigh of satisfaction. A good shed is so laden with experience and meaning that even a dog can sense it when it comes to the door, tail up and nose sniffing . . . A significant shed has good Feng Shed, a slight variation of Feng Shui, the Chinese art of arranging the physical world to bring good fortune.

A shed is a place to make things but it's also a place to park your story. And you can't just park your story anywhere.

I feel privileged that people have trusted me with their stories in this book and previous ones. I hope I have honoured those stories and those lives.

RIGHT Farm shed, Queensland.

MAKERS, BREAKERS AND FIXERS

Australians have always prided themselves on being a resourceful people, solving problems and repairing things with whatever few materials and tools are available. The fencing-wire repair, bush mechanic, 'making do', have-a-go, she'll-be-right approach to technology and engineering has been both a source of pride and a perceived national problem.

It's a source of pride because it comes from a tradition of resilience and survival under difficult conditions. It's cursed by captains of industry because it supposedly leads to a culture of shoddy work.

There's a difference between resourcefulness and invention. Invention is making new things, which may be things the world didn't know it needed: the

imperative is generally profit, not necessity. Resourcefulness is solving a problem with the means at your disposal. It's the difference between what you want and what you need.

In my previous books *Blokes & Sheds* and *Rare Trades*, examples of this resourcefulness pop up frequently, especially in the old bush skills and among

WHAT IS A BLOKE?

In common with a great many other Australian expressions, the origins of the word **bloke** lie somewhere in 19th-century London slang. Before that the trail gets a little cold and a bit funny. Some suggested derivations include Dutch **blok**, a fool, and Shelta (a gypsy language) **loke**, a man, but the best sounding one is Celtic **ploc**, a large stubborn person.

farmers. No-one has managed to bottle this elusive quality, but Queensland author Ron Edwards has done an excellent job of cataloguing traditional bush skills in his series of books on bushcraft (although much of his work refers to pre-industrial skills).

Part of this book is dedicated to examining the circumstances of resourceful problem solvers — their approaches, attitudes and experiences. Some are inventors. Some work in high-tech fields and some in more basic technologies. It will come as no surprise that a great many of them spend time in sheds of one sort or another.

Curiously, for a national characteristic that we set such store by, we haven't given much thought to the origins of the problem-solving strand of our culture.

Over the years a number of factors or imperatives have been important.

Isolation is clearly the biggest factor. For Aboriginal people before European settlement, culture and technology developed not only in physical isolation from the increasingly complex world beyond Australia, but almost completely unaware of it. The early European arrivals knew what was out there, but distance from Britain and its manufacturing power was crucial. A replacement part for a broken plough or machine could take two years to arrive, easily enough time to starve to death. So a way had to be found to make the broken thing work.

Poverty was also a great spur to solving problems. Australia in the early 19th century, before mining and agriculture became properly established, was a very poor country indeed.

Then there were the varied and unpredictable (at least for Europeans used to regular, orderly seasons) environmental catastrophes — drought, flood, fire, dust storms and insect plagues. The land itself was probably the biggest factor in moulding people's sense of needing to come up with their own solutions, even if they were just temporary ones. The options were often extreme, although the rewards for overcoming adversity might be great. By the end of the 19th century, Australians had used available technology to develop our mining and agricultural industries to a high degree and we were the richest country per capita in the world.

The 1930s Depression saw a fall in living standards for many Australians. The hardship and frugality of those times were etched deep in the psyches of the generations who lived through them. The memory is still strong for many older people, who cannot quite trust the continued abundance of the current era.

War, in particular World War Two, made a big contribution to this culture of resourcefulness. With the collapse of industrial supplies from Europe, Australian industry had to manufacture essentials that had been taken for granted. Whole new industries sprang up overnight, some successfully continuing after the war.

Since that time there has been more or less continued abundance of material goods. Our manufacturing sector has not been able to compete with cheaper goods, mainly from Asia, but there are many successful niche industries. It seems we no longer need the frugal resourcefulness that finds uses for unlikely objects.

That is, until we hit the question of climate change, a key message of which is that we are using too much of our resources too quickly and must moderate our habits of over-consumption.

Will climate change be the great new imperative for resourceful problem solving?

After all, the problems of climate change are physical problems, ones that public relations, politics and other forms of spin can't make a dent in. Solving them requires an understanding of materials and the properties of real things. These are exactly the sort of things that the people in this book grasp well, along with a proven ability to cut across disciplines, or to 'think outside the square' as the horrible cliché goes.

We need to remember that in most Third World countries this same sort of resourcefulness is still at work. There it is clearly understood that the materials and resources we so casually throw into the rubbish represent an enormous amount of embedded energy requiring only an agile mind and a deft pair of hands to turn it back into something useful.

The fact that Australians, despite living in abundance, still hang on to the culture of frugal resourcefulness, even if only in theory, is an intriguing curiosity.

How do we instil this sort of resourcefulness into the next generation? How do we embed it into our culture when most people have no direct necessity to use such skills in their daily lives?

In spite of the certainty that climate change will have an impact on us, there seem to be no clear answers on the horizon. The creativity of the people in this book offers a bright ray of hope.

WHO'S NOT IN HERE AND WHY

A number of things have changed in the ten years since the publication of the original *Blokes & Sheds*. The concern of a decade ago that backyard sheds would vanish has not come to pass, although there has been further rapid growth of McMansions, where Australians are living in bigger houses on smaller blocks of land, thus missing out on potential shed space.

Under the pressure of television lifestyle programs and design magazines, the backyard has become more inside than outside: it's been cleaned up and made into a 'personal statement' (the vanity of it all!), with cushions spread tastefully in a place where once there was a useful woodpile or a scattering of car parts.

With our huge houses, home cinemas and enormous, energy-sucking air conditioners, we are, tragically, becoming an interior nation. Are we more inward looking as a result? More selfish? We are certainly more isolated.

YOU CAN'T TURN THE BRAINBOX OFF WHEN YOU GO TO BED . . . IT HASN'T GOT A SWITCH ON IT, YOU KNOW.

RIGHT John Wake, inventor and mechanic (and Ralph Sarich's first boss).

At one stage I envisaged a book called *Sheds of the Rich and Famous*. This would have been a very slim volume indeed as the rich and famous, with honourable exceptions, are too busy being rich and famous to bother about their shed lives. The preliminary investigations were very depressing and I shelved that idea.

There are a number of interesting sheds that didn't get into this book — the ones used for illegal manufacture of automatic weapons, for instance, or the one belonging to World War One war gamers — the war gamers who have built a complete 1916 battlefield in their shed (with sound effects). They dress up several times a month, go down to their trench to listen to wind-up 78 rpm records and get drunk. People wouldn't understand, they said.

UNDERSTANDING MECHANICAL ADVANTAGE

Many of the people interviewed for this book expressed concern that most young people these days are simply not exposed to the fundamentals of machines. They don't mean the finer points of diesel motors, but the elementary machines that form the basis of most complex technology: the lever, the inclined plane and wedge, the pulley, the wheel and axle, and the screw. Together these form the basis of mechanical advantage. The crowbar, the shovel, the wheelbarrow, a pair of scissors or pliers, and even a cricket bat at work are examples of the lever.

What the resourceful problem solver does is see the underlying principles or combinations of them that make up the workings of most technology. The mental agility needed to do that is only gained by actually working with these machines.

Exposure to basic machines also gives a grasp of the principle of action and reaction — that action (or failure to take action) has consequences, sometimes life-threatening consequences. This failure to understand consequences is seen by interviewees as a direct result of the 'black box syndrome' — people don't know how an appliance works, as if it is an inscrutable black box, and if it's busted, they just replace it.

There is no longer the obvious mechanical link between action and reaction that a lever, for instance, would demonstrate clearly.

Modern children's apparent lack of interest in how things work amazes the older generation who acquired their survival skills through insatiable curiosity. To them life was a hugely exciting journey of discovery, for which the stimulation provided nowadays by the glowing screen is simply no match.

The fact that people seem more remote from the world, physically disconnected from it or blind to its power and richness makes many interviewees profoundly uneasy. If there is a certain amount of human experience that is necessary if we are to survive, a yawning gap may be developing in it.

Perhaps it's our own fault. As the German writer Max Frisch said, 'Technology is the knack of arranging the universe so we don't have to experience it.' All technology heads in the direction of less contact with the world.

The classical Western scientific/technological 'way of knowledge' that gave us the Industrial Revolution and dominated the world in the 18th, 19th and 20th centuries seems to have become unpopular in the countries that gave birth to it. Far fewer students in the West now study mathematics and the sciences, preferring to follow more lucrative careers. Or they seem more drawn to the 'inner wisdom' of the East (or E-Z DIY versions of it). In the meantime the countries that gave birth to those ancient forms of knowledge are now drawn to outer wisdom: India and China are training engineers at the rate of hundreds of thousands a year.

They must know a good thing when they're on to it.

THE ELEMENTARY MACHINES THAT FORM THE BASIS OF MOST COMPLEX TECHNOLOGY: THE LEVER, THE INCLINED PLANE AND WEDGE, THE PULLEY, THE WHEEL . . . TOGETHER THESE FORM THE BASIS OF MECHANICAL ADVANTAGE.

PROBLEM SOLVERS

Chapter 1
PROBLEM SOLVERS

There are any number of approaches to problem solving on offer. It's a major area of business activity and publishing, dominated by a certain E. De Bono and a troop of others spruiking surefire methods of solving problems with funny hats, boxes, squares to think both inside and outside of — you name it. It's also an area of massive and dense academic study by psychologists and others.

While I have categorised the people in this chapter as resourceful problem solvers, that label could be attached to nearly everyone in the book. Many have what is sometimes called 'tacit knowledge'. This is a term invented by scientist and philosopher Michael Polanyi, best summarised by his saying, 'We know more than we can tell.' As tacit knowledge is gained by personal experience, not from what is written down, by its nature it is not easy to communicate. A common example is the knowledge of how to ride a bicycle. No amount of reading can help you do it: only by experimental attempts, experience and ultimately a leap of blind faith can you take those first tentative journeys on a bike. Tacit knowledge is embedded in people's habits and circumstances, which are the subject of this chapter.

What do these people know? Although their practices and circumstances differ enormously, some common themes do emerge. These themes may seem blindingly obvious but they are perhaps rarer in the broad population than you might assume.

Be curious: A strong sense of curiosity and inquisitiveness is probably the most common quality of the people in this book, along with a certain mental agility. Ask dumb questions, but remember the answers. What is now termed 'lifelong learning' is something that many people here have adopted without having the name for it. They stay open to new ideas.

Talk about it: It may seem obvious, but talk about the problem. Lots of people aren't willing to do this simple thing because they're worried they might look ignorant or not so important.

Talking about something also has the beneficial effect of putting the idea 'outside your head'. You are forced to describe the problem in words, not as unspoken concepts in your brain. That act alone changes your notion of the problem. Several people referred to grasping the solution to a problem as the words came out of their mouths to describe the problem for the first time.

Look up a book: There's a wealth of knowledge out there. A public library is a beautiful thing. When all else fails, read the instructions. A thirst for reading about the world is just another form of curiosity.

Be observant and have your wits about you: 'Be awake to the world … You might learn something,' as one old farmer said.

Being observant and having your wits about you now sounds rather old-fashioned, a bit like reading Sherlock Holmes stories or being a boy scout. Perhaps it's because most people are now spoon-fed so much information they are as full as googs and have lost any curiosity or interest in what the world may offer the astute observer. As another old man, a tradesman, observed, 'The fella who is like a bit of blotting paper is very rare now.'

Draw it: It's extraordinary how often problem solvers use drawings. It can be the diagram on the side of some old roofing timbers or the notebooks of a rocket scientist such as Allan (p.18). It can be a precise technical drawing as commonly used in the metal trades, or the simplest of mudmaps. Drawings are a powerful thinking tool.

Shifting the problem from the brain to the back of an envelope, a beer coaster or a piece of plywood takes it out into the world — puts it a little more at arm's length from the brain. Sometimes just that shift can shed enough new light on the subject to show a solution. Sketching a problem with a pen or pencil is a different process from picturing it in the brain. Drawing the problem starts to put physical attributes to the idea.

The act of drawing a structure, a machine or a process is a step along the way to actually making it and getting a feel for how it will work. So you will often see someone planning to build a structure under load draw and redraw a curve or a shape, almost sensing the heft and strength of the idea in three dimensions as they are drawing it in two.

The act of drawing a problem can also bring more minds to bear on a problem. What was previously only verbally described (and so easily misunderstood) becomes more concrete — sometimes literally, as people draw on a cement floor with a piece of chalk.

Technical drawing, which has a whole set of rules for describing objects, is now rarely taught in schools, and drawing as a method of description and discussion now seems less acceptable than it once was. This may be because it looks like a scrappy work in progress in an age where marketing and verbal spin predominate. Many people are almost too embarrassed to draw a problem — better to sweep it under the carpet verbally. Making images and pictures is now the province of experts — the moviemakers, computer game designers and so on — and amateur efforts can be seen as embarrassingly shabby by comparison. This is a nonsense. The pleasure of drawing should be at everyone's fingertips, no matter how simple or complex the result.

The value of blunders: We live in a world increasingly intolerant of error. In conversation with problem solvers, it is remarkable how often the role of

mistakes and error comes up — not as a disaster but as an opportunity to learn something. Why doesn't this work? Why did that fail?

Because so many people don't understand the nuts and bolts of how things work, they have less understanding of what is possible and what is not.

It's another part of 'the black box' syndrome: 'I don't know how it works and I don't care as long as it does work.' People press the remote control button or the keyboard key and when it doesn't work, they are powerless.

So we expect — and even demand — that technology work 100% of the time.

Yet the history of engineering in the past 10,000 years shows that all technological change has come about through trial and error and that risk cannot

be completely eliminated. It's a question of degree and of using commonsense.

Apart from the fact that the risk cannot be completely eliminated, the attempt to do so limits the process of trial and error which enables things to be learnt and improvements to be made.

Just leave it: Time not only heals, it also solves problems. Patience is a problem solver as well as a virtue.

It's important to know when to leave a problem to stew. As Mark (see page 40) says: 'Sometimes you just have to walk away from something for a while. Otherwise you'll get frustrated and go the wrong way.' When time is not an issue it's a common way to go. People will leave an unsolved job somewhere prominent so they can come back and stare at it occasionally, 'as though magic beams are coming out of their eyeballs and working on it like science fiction,' as one witty type said.

A number of extraordinary people will deliberately 'forget' about a problem. John, an engineer and inventor, leaves problems for his subconscious to solve while he's asleep. He says, 'You can't turn the brainbox off when you go to bed — it hasn't got a switch, you know . . .' Amazing when you think about it.

I WORK OUT A PROBLEM BY THINKING ABOUT HOW TO GET AROUND IT, OVER IT OR UNDER IT.

Look past the paintwork: Probably the most practical piece of advice for the would-be problem solver is to get to know the underlying principles of how things work — to see past the paintwork. Experience is the best, if not the only, teacher of this skill, and the best place to learn it is in the shed.

Never give up: 'I work out a problem by thinking about how to get around it, over it or under it. I never give up.' These might seem like outrageously over-confident words but the habit of never giving up was common among the people interviewed: persistence backed up by supreme confidence. It didn't faze anyone that it might seem to contradict the idea of putting something to one side or leaving it alone.

Such confidence comes from having a breadth of experience — from being exposed to different ideas and methods, and from having solved other problems. Many of the other qualities on this list — curiosity, observation and knowledge accumulation — give a person confidence that he or she will eventually come out on top. There's also the imperative of survival: giving up in an extreme situation could mean dying of starvation or thirst. With enough experience of such circumstances, not giving up becomes a habit.

'I know there's an answer but I don't know what it is yet.'

GOOD SHED THINKING

An engineering factory that used a number of heavy metal presses had a problem: a 20 ton press needed to be moved into a space where there was only a couple of inches to spare at the back and sides and about a foot at the top.

The base of the press had to fit within a fraction of an inch into a slot in the floor and there was no space for rollers or any of the usual removable adjustment devices. The only logical option seemed to be to cut a hole in the roof, rerouting various services beforehand, and then lift the press in with the aid of a massive crane. It would mean thousands of dollars and a big drama but the brains trust in charge couldn't see any alternative.

As a casual aside, they asked the maintenance bloke if he had any ideas.

'Gimme a hundred bucks cash and I'll do it,' he said.

He'd fixed everything else he'd tackled, so there was no reason not to give him the money.

So he got the cash and went off in the work truck to return half an hour later with the materials to do the job ... successfully.

How did he do it?

He returned with a truckload of big blocks of ice and laid them out in a path from the open area where the press was resting to the spot the press had to be moved to.

Using pulleys, the press was lifted up onto the ice path and a small gang of people pushed it into place. Having minimal friction on ice, the heavy press was easy to shift and could be adjusted into the right fit as the ice slowly melted.

It wasn't lateral thinking. It was knowledge of the properties of things, or perhaps remembering how things were done in the past. Or just not wanting to see money wasted.

AS A MATTER OF FACT IT IS ROCKET SCIENCE

In a converted pig shed on the outskirts of Brisbane, Allan is examining the latest version of the HyShot scramjet, his hypersonic engine design which has put Australia at the forefront of this new form of propulsion.

As the project develops into a $70 million research program jointly funded by the University of Queensland, Australia's Defence Science and Technology Organisation and the US Air Force, it's interesting to look at Allan's background and his distinctly Australian approach to high-tech research.

Allan gained a Doctorate in Applied Mathematics and then moved into Mechanical Engineering, which provided an excellent foundation for the complex work needed to make an engine work successfully at eight times the speed of sound.

AS A MATTER OF FACT IT IS ROCKET SCIENCE

Underlying that foundation is a youth spent working in metal and making things — fixing cars, building boats and houses, doing jobbing work. This mix of broad traditional skills with advanced mathematics and engineering knowledge proved to be a powerful combination.

While he acknowledges that his early experience with traditional skills allows him to look at things from a different perspective, Allan emphasises that mathematics is the tool which provides the answers.

'Mathematics is a supreme art. It has a purity of aesthetic about it. But it's also a workhorse — it's useful and I like the crossover between the beauty of the mathematics and the final outcome.

'You have an idea based on gut instinct and you think it will work. You design and draw the idea, then calculate the loads and performance, energy, momentum, mass. If the results convince you and you're not breaking the laws of physics, then you know you can do it . . . after a lot of effort.'

In Allan's case years of effort can come to a head in a few short make-or-break seconds when the scramjet fires on top of a rocket falling from 300 kilometres up in space. Scramjets only start to work at hypersonic speeds (over five times the speed of sound). Data gathered during the short firing period is later analysed and used to develop the next model scramjet. Allan and his team are now working on the sixth such model. The first was assembled with a tiny budget and included parts purchased at the local hardware store. Even in recent versions some of the minor parts were purchased at a cheap auto parts store ('Why not? The car industry is quite advanced and works to good quality specs'), and family members do some of the machining in the shed.

In classic style the first HyShot scramjet was driven from Brisbane to Woomera Rocket Range in South Australia in the back of a ute, the engine resting on a bed of paperback novels.

While it's satisfying that Allan and his team were able to achieve a world first through such frugal means, the successful project is also a product of a distinctively Australian research outlook.

A certain lack of inhibition is one of our strengths, according to Allan.

'You throw ideas into the ring, and people come back at you. The hierarchical research approach taken in other countries doesn't normally allow that freedom to think and express ideas. The robust dialogue here forces you to develop a logical structure to your ideas.

'If the rest of the world doesn't think they can do a given engineering problem, they certainly don't think we can do it. But we say: why can't it be done? And we will not give up, no matter how much our backs are against the wall. And when there's no expectation that we will succeed, what have we got to lose?

'In other countries, there's also a huge fear of failure and a fear of retribution for that failure. There's a tolerance here with stuffing up, as long as something is learnt from it. The day you admit your mistakes is the day you progress.'

And then there's Allan's huge enthusiasm, a sort of passionate propulsion fuelled by exhilaration in the science and the scope of the endeavour.

THERE'S A TOLERANCE HERE WITH STUFFING UP, AS LONG AS SOMETHING IS LEARNT FROM IT. THE DAY YOU ADMIT YOUR MISTAKES IS THE DAY YOU PROGRESS.

In total, Allan and his team's ground-breaking work is a unique mixture of, on the one hand, systematic problem solving requiring rigorous scholarship, and on the other, intuition based on experience, imagination and resourcefulness.

No doubt as the project evolves it will lose some of the seat-of-the-pants qualities that endear it to anyone who has ever tinkered in the shed and dreamed. Already HyShot has become a more properly resourced project but we can hope it will never lose the enthusiasm and optimism that have got it this far.

'Time provides you with the ability to come up with answers. It gives you a chance to think. A lack of time can lead to a desperate but very good fix. When you're up against the wall, you strip away all the redundant stuff pretty quickly and get to the core of the problem . . .'

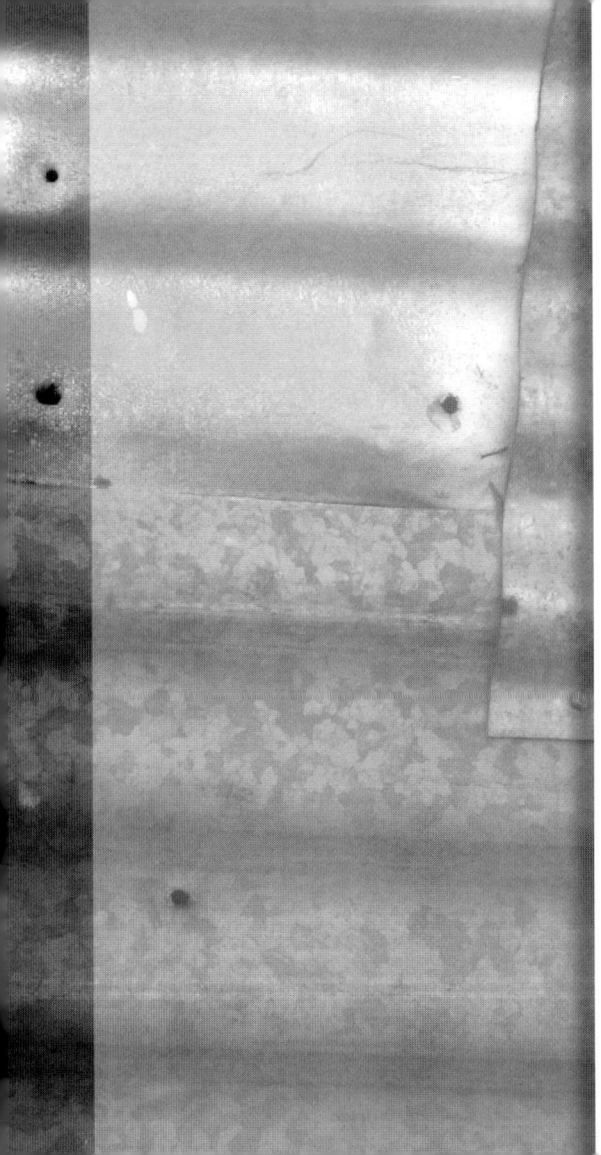

THREE GENERATIONS OF PROBLEM SOLVERS

What makes three generations of one family all unique problem solvers?

A good deal of it starts in Aynsley's shed, which he built from scratch 60 years ago when he and his wife bought a block of land on the outskirts of the city. Since then, several generations of the family have put the shed to good use.

Aynsley trained as a motor body builder in the days when mountain ash and Australian oak still played a significant part in the construction of cars and truck chassis.

Motor-body building at that time meant working in both wood and metal. Aynsley even learnt the now rare art of making wooden wheels. With a wide range of work experience under his belt, he eventually went out on his own and was able to tackle any fabrication job involving wood,

THREE GENERATIONS OF PROBLEM SOLVERS

metal or plastic, although he spent a lot of time constructing sheds and carports. 'I always worked in small businesses where you didn't send anything out to be done — you did it all yourself. You looked at a job or something to be made and thought, "I could do that."

'If I worked with somebody, I'd always be interested in how they did things. Observation is important. Textbooks — I get tangled up there, but if I see you make something then I can make it too.'

A few years back Aynsley worked on the reconstruction of the famous Birdsville mail truck, a 1936 Leyland Badger driven by Tom Kruse for twenty years in the 1930s, '40s and '50s. He also undertook a fundraising trip around Australia with an old mate to raise funds for the Crippled Children's Association on a couple of

50 cc Honda scooters supplied by Honda Australia. The bike is still in the shed, of course.

His son Terry works mainly as a specialist builder and fabricator, not only of houses but of large-scale sculptures and artworks. He and his brother helped fabricate a 30 metre high human body in steel and concrete in London several years ago.

'These jobs involve a combination of structural engineering decisions with aesthetic judgements. I like things that have a practical and aesthetic solution. Often that happens to be the simplest solution too.'

Terry's ability to figure out a good solution gets him work that other people don't want — or don't know how to tackle — such as building stairs or

fixing vertically-hung sash windows. The solution usually requires an understanding of structure combined with aesthetic considerations to arrive at a workable form.

Understanding structure is vital to problem solving according to Terry.

'First you have to understand the structure of things. You don't go hacking into something without looking at it very carefully to see how it's made. If you don't look you'll probably end up ruining your saw blade on a nail you didn't see or didn't expect to be there. Or worse still, you end up compromising the safety of the job.'

According to Terry, the family tradition of problem solving is about exposure and access. 'If you put a hacksaw in someone's hand, they wouldn't know how to use it properly unless they were told. So unless the hacksaw is there to experiment with, you'll never know anything of what it could do. In my case there was always an abundance of tools and materials which my brothers and I were encouraged to use — it was virtually expected.

'The same thing applies to my son Sean — but he grew up with a metal lathe and mill in the shed.

'Sean and I are both slightly dyslexic but we can both visualise things pretty well from a verbal description. Some people simply cannot visualise.'

Sean also had the now rare benefit of growing up with Meccano and Lego Technics (a kind of motorised plastic Meccano). This gave him a good grasp of fundamental mechanical principles. In Sean's Year 8 science class the teacher gave groups the task of making a simple two-speed gearbox from mechanical Lego. Fifteen minutes later, Sean presented a surprised teacher with a four-speed gearbox that worked flawlessly.

Sean: 'I was really lucky having a grandpa who could do everything in wood and metal. I learnt a lot from him. And my other grandpa was an electrical engineer whom I spent a lot of time with, so I learnt plenty off him too.'

There's a family story that Sean at a very young age built a tiny set of operating traffic lights for snails in the garden.

Sean now works as a pattern-maker in the foundry and casting field, a traditional manufacturing industry that is undergoing rapid technological change. There is a strong history of problem solving in pattern-making, which is partly the attraction of the job for Sean. His hobby is making rockets, and his high-end metal-working skills are useful for making precise high-temperature nozzles, venturis, gas jets and connectors.

Aynsley's shed is still a place of excited learning. When his youngest grandchildren come to stay, they want to do everything. 'They get out of bed early and rush down to the shed. One of them is learning woodturning on the lathe. He's pretty good too.'

YOU'RE NEVER TOO OLD TO PLAY

Years back, in a small town on the River Murray, you could visit the most interesting playground you are ever likely to see. There were slippery dips, flying foxes, immense rocking horses, strange seesaws that didn't seem to make sense, rotary cones, roller coasters, wavy and bumpy slides, an instant mud machine — 180 items altogether. Maybe it had a certain element of risk, but nearly everyone felt that just made it more exciting.

The 180 items also provided informal lessons in Isaac Newton's three laws of motion, as kids — most of them under the age of 60 — tested the boundaries of momentum, force and velocity. There might have been a scraped knee or two in the course of this scientific investigation. However, a few injuries, talked up by the occasional legal opportunist and nervous insurance company, eventually put paid to the Monash playground as the best playground in the world.

YOU'RE NEVER TOO OLD TO PLAY

The most dangerous (and interesting) equipment was removed and taken back to where it came from — to Grant Telfer's shed company.

Grant was a 'blocky kid'. He grew up in the Depression on a bush fruit block, where his family had little money but an abundance of two useful commodities: time and space.

As a child, Grant was in and out of his grandfather's coachbuilding works, where he picked up all sorts of skills. With help from an electrician uncle, at 12 or 13 years old he built himself an arc welder — at the time, a new and exciting form of technology.

'Whenever I turned on the welder it drained so much power from the local power supply that the neighbours' lights would dim. I soon found uses for the welder. One in ten of my projects worked properly but the others were never a complete loss — they all worked somehow.'

At that age Grant was an enthusiastic reader, as he still is to this day. He studied in the city to become an electrical engineer but found the work, especially the maths, difficult. So he returned to the block. Among other things, he joined the local gliding club. When the club needed a new hangar, Grant rose to the challenge. He based his design on a tubular steel building he had seen, principally because tubular steel was the material available to him. By hammering flat the ends of the tubes and using some angle steel, Grant came up with a strong, effective truss design that turned out to work well for sheds and farm buildings.

Grant soon had a good little business going. His buildings were popular with farmers because they were delivered unassembled and modestly priced. At the time there were no regulations requiring detailed engineering specifications for a farm shed, and Grant's designs were pretty much intuitive.

Grant is a strong believer in intuition based on experience and looking. 'It has to look right. If you go the way your talent leads you, you'll go a lot better than if you are forced to do something that's not you. The intuitive part comes first and the logical part checks it out.

'Engineers can easily design something for you that is over-engineered to the point of being uncompetitive. I had to find people that thought my way — with a certain frugality.'

Grant's designs evolved over the years. The introduction of steel C-section structures to replace timber purlins and other structural members required a change in technology. 'Instead of so much welding, our suppliers roll galvanised C channels to shed lot lengths and punchings, supply bolts to suit, and even deliver to the factory for minor final work. Where there were 15 people in the factory, there are now two. You become a designer/marketer rather than a manufacturer.'

Though the business was successful, it had its ups and downs from droughts and so on. In slack periods Grant would make playground equipment. As president of the local Institute, he knew there were five acres of unused saltbush in the town. He put some of his playground equipment

there among the saltbush, and people came. So did more and more equipment . . . until the five acres were full. Designing the equipment was itself a form of play for Grant and his co-workers, who all got a kick out of making a challenging playground.

BELOW Monash playground in its heyday. An equal number of children and adults waited for a turn on the slippery dip.

The Renmark Pioneer

The playground became nationally famous. One person writing from the US described it as being better than Disneyland because people had to work at it a bit to enjoy it.

Grant has retired from the business now but keeps busy, still working on powered bicycles and other ideas.

What did people gain from their experiences at the playground?

'They got a great sense of fun and satisfaction. Vandals loved it — and I loved them, because they were unpaid testing teams. There was a great sense of wonder at all these things people could not have imagined existing anywhere, let alone in a small country village.'

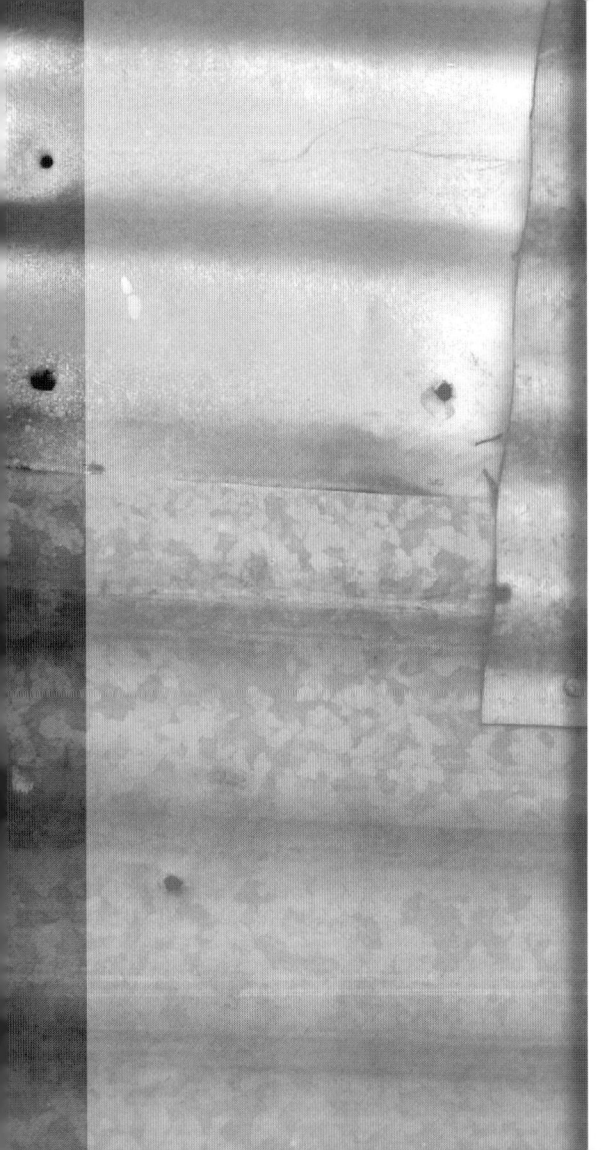

GOING BANANAS

'If you can survive on the streets of Cairo, you can survive anywhere,'
says Ramy. 'It's big — 30 million people — crowded and in parts
impoverished, so it's very competitive. By the time I was eight years old,
I could at least survive the streets.'

As a young man in Egypt, one of the ways Ramy found to make money
was to sell handmade papyrus, usually to European tourists. The mental
agility acquired on the streets of Cairo serves him well in the project he
is engaged upon now — mechanising the old papyrus-making technology.
Engineering training he has received in the interim in Germany and
Australia means he has a good balance between systematic and organic
approaches to problem solving. Ramy acknowledges that he learnt plenty
from his illiterate grandmother: 'She had 10,000 years of wisdom and

IT'S NOT AN EASY JOB . . . IT TOOK YEARS OF EXPERIMENTING TO GET IT RIGHT.

understanding behind her, with a profound respect for nature and humanity. That's not to be dismissed lightly.'

It's a long way from the hot streets of Cairo to an even hotter shed in Australia where he and his mate John are making adjustments to a machine that turns banana trunks into sheets of papyrus.

'It's a cheap, sustainable, biodegradable product that has the potential to fill a huge niche. Unlike paper it won't split when it gets wet and it doesn't have the environmental drawbacks of plastic bags. It can be made thick, thin or laminated like plywood and it can be coloured. And it's made from an easily grown raw material that's not used for anything much right now.'

To tackle the task of mechanising papyrus-making from banana trunks, enter John. Ramy had heard about this all-round engineering bloke who could make things with his hands and asked him to help make the machinery he needed. Like any self-respecting shed scientist, John rose to the challenge.

'It's not an easy job,' says John. 'The banana trunks are wet, of course, and if you turned them on a lathe the way you do to make plywood, they'd just come apart. It took years of experimenting to get it right.

'The first run of papyrus that we made happened by accident. We'd given up. It was seven o'clock at night, dark and wintry. We turned the machine off and it made a different sound. We looked back and there it was — as the machine slowed down it was making separated banana.'

'Slowing down is sometimes the best way to get a result,' says Ramy.

Some ten years after they started working on the idea, the problem is now virtually solved.

Ramy distinguishes their kind of invention and engineering development from traditional bush mechanics. He and John have all the technology and engineering solutions they could need at their fingertips, borrowing widely from printing, plywood-making and textile industry technology — it's not a solution found on the run from the limited resources that happen to be at hand.

A quantum leap still has to be made at some point. Ramy is keen to emphasise that his project is about making something new, not an incremental market development of the kind sometimes passed off as a 'breakthrough' — a smaller DVD player, a room deodoriser that goes beep, or any of the thousands of 'new' consumer goods foisted on us by junk-mail catalogues.

THOSE THINGS USUALLY COMBINE EXISTING IDEAS AND ARE NOT REALLY INVENTIONS AT ALL. THEY DIVERT RESEARCH AND DEVELOPMENT FUNDS AWAY FROM PROPER INNOVATIVE ENGINEERING AND SCIENCE. THEY ARE ALSO ABOUT PROVIDING A WANT, NOT A NEED.

'Those things usually combine existing ideas and are not really inventions at all. They divert research and development funds away from proper innovative engineering and science. They are also about providing a want, not a need. We need paper, as we need water, fresh air and other necessities.'

Given that each of them comes from a place where the imperatives to survive are clear, it's not at all surprising that the apparently odd combination of the German-trained Egyptian engineer and the farm-shed-trained bush mechanic hit it off so well.

YOU BUST IT, I FIX IT

That's what it says on Noel's business card. The card lists an impressive array of things he'll fix — motorbikes, magnetos, musical instruments, radios and clocks — and offers lathe and milling machine work.

A short walk from one chockablock room to another in Noel's network of sheds proves that he does indeed tackle all these things and more.

On one bench lies a carefully disembowelled 1950s Wurlitzer jukebox. Noel points out the Wurlitzer's solid mechanisms, which predate electronically controlled technology. 'I like working on things like this because you can see how they work, or at least deduce it. You can see something happening. I know the reliability of the old gear is probably just fair, but there's nothing to see with the new stuff. With something like this jukebox, all the insides are on the outside.'

PROBLEM SOLVERS
YOU BUST IT, I FIX IT

Noel is typical of the local problem solver. People front up to him with all manner of mechanical and technical challenges. Most of the time he takes them on, even though his formal training in any of the relevant subjects is limited.

Apart from the usual strategy of consulting manuals and books, Noel's secret lies largely in his connection to a wide network of fixers and repairers who share knowledge. Such networks can be semi-formal, as in clubs of motorbike, vintage-car or old-machinery enthusiasts, or they can be completely informal, based on the members' wide experience and knowledge of who can do what. Far from being isolated, people like Noel have a very social way of living and working in a shed, with lots of visitors and people needing repairs.

The question of sharing knowledge is an interesting one: 'If they establish that they have a genuine need to know something, then I'm happy to help. I've benefited a lot from the old blokes telling me things.' But information is not given out willy-nilly. 'Sometimes you get time wasters, people you know have no intention of actually doing anything. If they're the kind that will hop in and do a project, then I'll help them.'

Noel spends a fair bit of time informally training young people in skills such as welding and mechanics. 'I show them how an oxy works for five minutes then let them have a go. Some pick it up, others don't. The ones that do pick it up are worth helping. The rest are too lazy and just sit in front of the television or computer.

'Funnily enough, I find that girls make some of the best mechanics. They're not so inclined to pick up a hammer to solve a problem — they'll stop and think. They'll use their brains first and are not so short tempered.'

What distinguishes Noel and his colleagues in the general-repair caper is a broad and never-ending interest in everything that opens and shuts, makes a noise or a note, has a wheel or gears.

Unending curiosity is the secret to a good shed. And, yes, that's a copy of a Morgan three-wheeler Super Sports that he's standing next to. 'I couldn't afford an original so I made my own version. People seem to find the copy just as interesting. It's not bad for an all-up cost of $100.

CANNIBAL HONOUR

The art of cannibalising and recycling comes down to perception.

When the average all-consuming citizen looks at a piece of broken machinery, a non-functional tool or a discarded household object, they see a worthless piece of junk best slung into the rubbish.

The accomplished cannibaliser/fixer/repairer looks at the same thing and sees a set of machines designed for mechanical advantage and, for the most part, still in good working order. They also see components that required a great deal of energy and effort to make. They see the elemental properties of the thing, not the brand, the colour or the styling.

Take the ordinary everyday bicycle. A bicycle is a marvel of mechanical efficiency. The chain, cogs and gears amount to a simple and almost faultless device that can be used to drive almost anything. Years ago, Australian shearers (who were great users of bicycles) used a modified stationary bicycle to drive a grinding wheel to sharpen their shears. In Third World countries, the bicycle is used to drive all sorts of water pumps, mills and generators. It makes a lot of sense to use the biggest muscles in the human body — the legs — to do such tasks.

And there's more to the bicycle than the chain, cogs and gears. The wheels, which run on very efficient and cheap bearings, can be used as the basis for fans and wind generators, mounted vertically or horizontally. There's the story that Australian prisoners of war in a Japanese camp built a centrifugal plasma separator from a pedal operated bicycle wheel mounted horizontally. (Weary Dunlop, the famous doctor who treated so many POWs, attributed the Australians' high rate of survival to the fact that many of their number were from the bush and knew how to survive.)

A bicycle wheel, connected to a smaller cog or gear, makes an excellent reduction gear to turn the fast but low power from a small motor into slower, greater power (so that a small motor can be used to turn a spit, for example). The rubber from a tyre tube has dozens of uses — as a belt, a rubber band, a slingshot. The spokes, usually made of good quality steel, can serve as hooks, skewers or tools.

All that usefulness from something slung out in the hard rubbish collection because someone hasn't bothered to replace the brake pads or (unbelievable but true!) because it has a flat tyre.

It's not just bicycles that are good resources for the recycling cannibaliser. Cars also represent a host of opportunities. Windscreen wiper motors are an example: small 12-volt motors with a readymade reduction gearbox. White goods also have many useable parts: not just robust motors and excellent gearboxes, but also drums that can make excellent plant pots, fire boxes or barbecues.

Truck and car differentials can take on whole new lives of their own thanks to their ability to reduce or increase power and speed or to change its direction in a right angle. They are to be found in any number of cobbled-together opening, closing, winding and lifting mechanisms.

Even a basic object such as a hacksaw blade need not be thrown out. Many leatherworkers and other artisans treasure knives made from old hacksaw blades, whose high quality hard steel holds a good edge.

Almost any commonly used object that is made from good materials or has useful mechanical principles underlying its construction is ripe for reuse.

Transforming what is considered waste back into useful articles is an immensely creative process but it is not as honoured as

PERHAPS IT CAN BE SEEN AS A SORT OF MECHANICAL JAZZ, A WILD IMPROVISATION THAT PROVIDES AS MUCH CREATIVE PLEASURE AS ANY MUSICAL JOURNEY.

other kinds of creativity. The cannibaliser is considered 'clever' in a quaint and rather grubby fashion but, perhaps because their work is not easily commodified (that is to say, money can't be made from it), it is all too often dismissed as old-fashioned or even sentimental. The logic is something like this: because nothing will ever be scarce again, why should we bother with old stuff?

There are no textbooks for the scavenging bush mechanic and, unless you count a few bright people in sheds, no courses for the would-be cannibaliser. It's a genuine oral culture, one of the few genuine ones we have, and it's on the verge of becoming lost wisdom. The engineering knowledge that gave the Industrial Revolution its tremendous power and laid the foundations of our current prosperity is now in danger of becoming invisible and forgotten.

Perhaps that lack of formal endorsement is part of the attraction of this oily-clothed, steam-powered wisdom. It has become a Western way to esoteric knowledge that can be attained only through long studies with the shifting spanner and the holy angle grinder. Or perhaps it can be seen as a sort of mechanical jazz, a wild improvisation that provides as much creative pleasure as any musical journey.

Resourceful re-use is an island of practical commonsense in the roaring flood of consumer advertising that washes over us constantly. Unless we are buying things in glossy packages with the smell and look of novelty about them, we are told, we are not alive. Re-used and repaired

ABOVE Some bike and washing-machine parts, a 44-gallon drum: hey presto, a cement mixer.

objects do not glitter enticingly or offer a fabulously glamorous lifestyle (whatever that is). In fact, they remind us of dark days of scarcity.

The point is: will we ever again live in times of resource scarcity? Will we need once more to see beyond the surface

and look at things for their genuine use and qualities? And if we do, will there be enough people around like those in this book to make sure we find a way through?

One contributor to this book probably spoke for many when he said that he suspected that our current world was like a light bulb, burning brilliantly but fragile and brittle in itself. Everything could change easily: every power failure is a sharp reminder of how dependent we are on complex interlocking systems.

Perhaps that's too pessimistic. Perhaps we really are a deeply resourceful culture at heart and could cope with massive sudden change or collapse. You could always look on the bright side: one optimist pointed out that the upside of people throwing things away is that there is lots of good stuff at the dump.

GUESS WHOSE DAD'S GOT A WHIRLPOOL (OR 25)

'This shed was built on a budget of stuff-all. There was not much money from working in the timber industry — still isn't — and because we couldn't afford a lot of the machinery I wanted, I built that too.'

Mark has built a range of perfectly functional wood-working and metal-working tools such as sanders, bandsaws, lathes and drill presses from dismantled white goods.

'Old Malley's Whirlpools are very useful. I suppose I've dismantled about 25 of them from the local dump or wherever. They've got good half-horsepower two-speed motors and the bushes, bearings and shafts are useful because they are all a standard size, 5/8 inch. This means that I can build machinery from the parts — all the parts are interchangeable.' The cast wheels which the washing machine's drum rests on are useful

GUESS WHOSE DAD'S GOT A WHIRLPOOL

too: Mark's bandsaw blade runs on them at the top and bottom. The washing-machine lids become protective side panels for the bandsaw — being hinged, they give easy access to the blade. Washing-machine gearboxes can be used to change machine speed ratios for various jobs.

Mark is proud of his Aboriginal heritage, but his work as a ranger and educator in the local Aboriginal community brings a certain amount of pressure.

'You tend to be working with the head rather than the back. So when I finish work I usually make my way out to the shed. It's not too bureaucratic here.

'This shed was a chook shed about 15 years ago. When we moved into this house, I stored a stack of beautiful timber in the old chook shed. I went up there one day when it was raining, and it was wetter inside than out, so I built a bigger shed over it and it's grown from there.

'I've always liked making things. I was always the kid who was building billycarts. I loved science too, the logic of practical things. I learnt a lot off my dad, and my granddad, who was an initiated clever man. I spent a lot of time with him. He could survive anywhere, build anything.'

Like his forebears, Mark has spent a lot of time working in local timber mills.

'You learn a lot from rough-cutting timber and being around people who have worked with timber all their lives. I like turning wood on the lathe and making things with wood. We're lucky around here because there are lots of camphor laurel trees and I can generally get the wood for free.'

The lathe that Mark uses is, of course, home made. The heart of his shed is a work space where his power tools are stored in a number of old poker-machine bodies ('They're pretty strong and easy to lock up'). A huge collection of bushes, bearings, shafts, nuts, bolts, screws, washers and springs are kept in a storage system built out of recycled bed frames and oven trays: 'Each tray is labelled and all the parts are easily accessible.

'I've kept anything that's mechanical or has a potential use. As soon as I bring back something from the dump, it gets dismantled, categorised and filed. The system is very efficient.' Old Eskies are used for storage, as are the metal vegetable crispers from old fridges — they are termite resistant. It's resourcefulness at its best: nothing is wasted.

Although Mark is kept busy with his community's cultural activities, he plans to retire to the shed permanently as soon as possible.

'I have to spend time here regularly to have a good grasp of where everything is. I also spend time with my boys here — sometimes three of them are here making things. I've been teaching them all the trades I know because I think they should be able to fix anything.

'As a Gumbaingirr Aboriginal man I utilise the resources I find around me, as Aboriginal people have always done. Before 1788 we used what was naturally available. Since then my forebears and I have learnt to use both what is naturally available (such as the wood I use in my woodwork) and the rubbish discarded by non-Aboriginal people. I take it apart,

compartmentalise and store it until I build from it the stuff I need to teach my boys how to use the resources around them. I guess I am able to see the value of all resources available to us — another man's trash I can turn to treasure.' ▤

BELOW Mark and his three sons — shed geniuses in the making.

NEAR ENOUGH IS NEVER GOOD ENOUGH

When his father died, Toby was the eldest child in his family. Their cattle property west of Ingham in North Queensland still needed to be kept running so Toby, at the age of 11, took over the simple vehicle maintenance and kept the 32-volt lighting plant and the water pumps going.

Toby tells this story now in a matter of fact way. He had been helping his father for years. He knew how everything worked, and he'd been driving since he was nine years old. At the age of 13, he was granted a special driving permit for the property's old Ford Blitz trucks.

Toby followed up his avid interest in engineering and mechanics by working in garages, farms and stations around Queensland. He was always learning and absorbing information like a sponge. 'When I was a kid I read *Dyke's Automotive Engineering Manual* from cover to cover,'

NEAR ENOUGH IS NEVER GOOD ENOUGH

he says. 'That gave me the basics. But I've always tried to read something every day — usually a technical or workshop manual — that will help me understand more about how things work.'

Like many other multi-skilled people, Toby has never even thought about stopping learning. Decades of learning makes for one very resourceful problem solver, so Toby was eventually able to set up a field service business operating mainly in north-western Queensland.

His diverse skills meant that he could pull up to a property or station and fix anything from a non-functional sewing machine to a busted D9 bulldozer.

'There's a shortage of good tradesmen out in the isolated areas. I loved doing that work. Anything in the mechanical/refrigeration/electrical area — generator sets for example. Every time I got a new job I enjoyed the challenge of working out precisely the cause of the failure and rebuilding from there.'

To be able to pull up to an isolated location and repair virtually anything requires some unique skills.

'It's impossible to carry all the tools you could need. I carried one and a half tons of them on the Landcruiser and it still wasn't enough. So I carried tools for making tools, such as a converted Hercus lathe. When the nearest workshop could be a 600-kilometre round trip away, you have to be able to make tools on the spot.'

To illustrate this point, Toby pulls out a very large socket spanner. 'I made this to work on the back end of a 1942 Caterpillar D8 . . . I used some scrap steel and the 1-inch cutting edge off an earthmoving scraper.'

What is striking about the socket spanner is the quality of its finish: it looks for all the world like a factory-made tool. There is nothing slapdash or 'she'll be right' about this piece of 'bush-mechanic' toolmaking.

'Near enough is never good enough. I'm very fussy about precision and always worked to at least one thou [one thousandth of an inch]. I had to work to those limits: if a job is worth doing, it's worth doing well.'

Fussiness about precision doesn't stop Toby salvaging found objects. The crane on his trailer (shown in the photo) is made up of found bits and pieces — a crank on the end of a Holden axle turns a gearbox from a 1930s Rugby 6 and easily lifts an engine block. The trailer rests on a wind-up stand made from a piece of nicely threaded steel he found on the side of the road. 'It was too good to throw away and I soon found a use for it.'

Toby's retired now but only eight years ago, at the age of 60, he finally bit the bullet and, for insurance purposes, got a formal qualification as a mechanic. Not surprisingly, he passed the oral examination with flying colours.

Toby now works in his shed on the Atherton Tablelands and is part of that admirable network of people who restore old stationary and farm engines.

The old farm pumps and engines may seem like crude old world technology but they are part of a can-do approach to technology that is anything but crude.

'I loved the work I did, but it was serious work. You had to be thorough, you had to be accurate, you had to get the details right or there could be serious consequences. For instance, a split pin not put in properly can lead to a disaster in some isolated spot on a rough bush track. How you do your job matters.' ≣

BELOW The socket spanner Toby made on the spot from an old grader blade.

THE LOCAL FIX-IT BLOKE

Living in a small regional town has both benefits and drawbacks. If you're someone like Terry, you have a sort of legend status as the local fix-it bloke, a useful resource that is not sufficiently acknowledged. The drawback, Terry says, is that the local council's been trying to close down his back-street shed and yard for the last 30 years.

'I like my toys, you know,' says Terry. 'They're the real thing.' He's referring to his collection of Chamberlain tractors and other earthmoving machinery to be found in various states of undress in the yard around his shed. The Chamberlain tractor is a famously robust Australian-made tractor that went into production immediately after World War Two and was built for nearly 40 years. Around his shed and yard are to be found many other icons of Australian-made technology

THE LOCAL FIX-IT BLOKE

such as Dunlite wind generators and old radios with brands like AWA, Kreisler and Ferris.

Terry's a passionate believer in Australian-made technology and is keen to preserve as much of that tradition as possible. He has a main shed in which he spends a good deal of time: it has the obligatory comfy chair and a big collection of technical manuals. What makes Terry slightly unusual in the shed-science department is that he's a double threat: he's both an electronic technician and a mechanic. He's got a lot of spare parts and yet-to-be-cannibalised gear spread through several sheds. Some of the sheds are so crowded that he has to jump over things to get in or out.

'I know roughly where things are,' he says. 'I still mutilate massive amounts of things. People seem to have plenty of money nowadays so they don't need to dismantle things for parts any more. And some modern electronic components are impossible to dismantle — such as "surface mounted devices" [smaller electronic components that are attached on only one side of a printed circuit board]. Once you could take electronics apart easily and reassemble them. These are so small that just the process of removing them can break them. There's no fun in it any more.'

Although his shed may have the slight illusion of disorder, Terry likes to maintain a certain decorum. He's famous for wearing a tie and has been doing it for 45 years.

He's obviously a good source of knowledge for farmers and machinery owners in his area and always has lots to do and play. He can't understand how kids don't get bored with gazing at television and computers all day.

'As you get older, you do weirder things. Now that I'm over 50, I don't feel the need to conform. At this age, you've achieved what you're going to so you might as well do what you want to do. I like to test theories, even if they're a little bit dangerous. You don't want to get bored.'

I STILL MUTILATE MASSIVE AMOUNTS OF THINGS. PEOPLE SEEM TO HAVE PLENTY OF MONEY NOWADAYS SO THEY DON'T NEED TO DISMANTLE THINGS FOR PARTS ANY MORE.

THE UNDERGROUND SHED

If ever there was a place where resourceful fixers, menders and improvisers gather in big numbers, it's Coober Pedy in the flat, dry country of South Australia's outback — the most successful opal mine in the world.

Coober Pedy is a place of myth and legend, populated by the adventurous, the resourceful, those on the run from anything and everything, and the slightly mad. Everyone has a nickname and everyone has a story to tell. Along with the absence of anybody to say 'It can't be done', its isolation has been the driver of some truly remarkable mining innovations.

Many Coober Pedians live underground — a smart move in a place where the mercury sits above 40 degrees for most of the summer months. Underground 'dugouts' are mostly comfortable, cool and, thanks to the tunnelling machine, reasonably cheap.

The tunnelling machine, a tracked-crawler vehicle mounted with revolving stone-cutting heads, is one of a number of engineering innovations more or less unique to Coober Pedy.

Probably the best known of these innovations is the blower, a kind of vacuum cleaner mounted on the back of a truck that sucks up dust and small rocks from underground, clearing the mining area much more rapidly than the previous shovel-and-bucket operations. The Coober Pedy landscape is dotted with all manner of blowers cobbled together from old diesel engines, cooling ducts, industrial fans and other bits and pieces. From the first one invented about 40 years ago they have evolved into surprisingly efficient devices.

ABOVE Looking like strange giant insects on the backs of trucks, mining blowers are just the tip of the resourceful devices made in Coober Pedy.

ABOVE Noodling machine at work: the tiny shed with the chimney is where the opal is checked with the UV light.
RIGHT Otto, one of Coober Pedy's innovators.
FAR RIGHT Coober Pedy underground workshop — or a tiny part of it at least.

Another local innovation is the noodling machine. This was devised to address the fact that the white cones of mullock (waste material) scattered across the landscape still contain opal that was missed in the original mining operation. Noodling originally referred to the slow manual process of sifting through the piles of mullock with a shovel and sieve.

Opal fluoresces under ultraviolet light (black light). A noodling machine crushes the mullock into small pieces then runs it on a conveyor belt through a small darkened room lit by black light. Any opal in the mullock will flash under the light, and a watching miner will grab it from the conveyor belt. It's a highly addictive activity as at any moment a precious lump of opal could suddenly appear.

A typical noodling machine like the one in the photograph is an odd assortment of machinery, often mounted on a trailer, the back of a truck or even an old bus. The first mechanised noodling operations were brilliantly successful and a great deal of the mullock that has piled up over half a century of mining has now been mechanically noodled.

There are many other examples of locally made mining technology — drilling rigs, augers and cranes — all assembled by astute local engineers.

Although there are a number of well-known local engineering businesses that operate in conventional workshops/sheds, there are also the lone innovators such as Otto, whose underground workshop is shown here. Underground workshops like this can be misleading. What may appear to be a small space crammed with gear will often have a discrete tunnel leading off from a far corner. That tunnel will have another leading off it and so on, and all of them full of gear that will one day come in handy. Many opal miners go to auctions and clearing sales of old machinery businesses and buy up stock to hoard in their tunnels.

Otto has built some of the most impressive equipment around Coober Pedy, well thought out and well executed.

Although opal seems to have reached a plateau in terms of sales and availability, its miners remain an unrestrained force of improvisation and seat-of-the-pants problem solving.

RIGHT Otto's sophisticated blower and bucket chain device.
BELOW One of the many blowers and noodling machines dotted about the desert.

THE GOOD SPACE

To me this shed is like the journeyman coming home and building his masterpiece.

'There's a European tradition going right back to the days of the guilds that once a tradesman had done his apprenticeship he went out into the world as a journeyman and broadened his knowledge. He would eventually return with all that knowledge and build something — his masterpiece — that showed everything he had learnt.

'I never did a formal apprenticeship but I think I've served my time. I've built boats, designed and built several houses, worked as a builder and cabinet maker in Scandinavia and other parts of Europe, done any number of renovation jobs.

'This shed was consciously made to be a good environment to be in.

THE GOOD SPACE

The doors open up so it's cool in summer. It's got a nice high roof so the heat doesn't press down on you in the hot weather. It's got big windows for good light and there's enough room for me to build a proper boat, which is the next big job I want to do. So it's a nice place, a good space to be in.

'Working in the building trade, you make decisions about what you want to do. My father was a very good builder but wasn't good at whipping people . . . and neither am I. That's what you have to do in the building trade if you want to make a lot of money. Or you can forget the money and decide to be good at the trade. That's what I decided to do.

'I know from working both here and overseas that the Australian building industry is very efficient at knocking out square metres of house. A building framer can erect an entire wooden house in a few days. But design is a problem. There are so many imitations and rehashes of styles from other times and places — Spanish haciendas and English cottages and other nonsense.

'We need to be ourselves, to stop copying other people and build for our own time and our own place. Grow the verandahs with deciduous vines to shade the house in the summer and let in the winter sun, save all the water we can with big tanks. It's about the right technology for the climate.

'Most people want to build castles instead of shelters. A castle doesn't have a roof. And here the roof is the most important part of a house. What you clad the walls with is irrelevant. You've got to have a big hat for your house. That's why a style of house like the traditional Queenslander works so well — it has a sort of integrity to it, coming from the influences of Indian houses and so on.

'I don't know if there's a particular problem-solving frame of mind that goes with building in Australia but the people I see, who are good problem solvers in the building industry, are very thorough — they get the geometry right, they do lots of preparation, they get the sequence of building events right. Building well is not simple. You can see that from the way people use silicon these days to fix everything. Silicon is a wonderful thing but in the past people built houses that didn't leak — because they got the preparation right and they used materials correctly, with integrity.

'Paying properly for time, quality of work and good materials is not a message people want to hear, but then again as my dad used to say if you want good advice about building, ask a butcher because he will usually agree with whatever you say ...'

RIGHT Like so many other shed users, Chris likes to draw out planned jobs on the floor, preferably to real size, to help nut out a problem.

Chapter 2

THE FARM SHED

THE FARM SHED

The farm shed remains the original resourceful shed. A 19th-century Australian farmer confronted with a broken implement could not wait around for a new part to arrive from Birmingham. He would fix it on the spot if he could, or take it to the local blacksmith's.

Because farming has been more vulnerable than other occupations to the fluctuations of a fickle climate, farmers have also had to deal with challenges and opportunities from that quarter. As a result they would often accumulate large stockpiles of old equipment and materials, with which they could meet any situation or make just about anything.

Farmers are no longer so physically isolated, whether in relation to resources (most spare parts can be delivered within a day or two) or to information and knowledge (weather forecasts, commodity prices and so on are now instantly available online), and they now use a wide range of computer-based technology such as GPS systems and the like.

Despite this massive technological change, many farmers are not keen to give up their old stockpiles out the back, even as scrap-metal prices rise to great heights.

The availability of cheaper machine tools such as lathes and mills means that many farmers now have sophisticated engineering capability at their fingertips, so the habit of adapting and improving machinery has become well and truly ingrained.

The drive to modify and improve existing farm machinery has a long history. From Bowyer Smith's stumpjump plough and McKay's Sunshine stripper harvester to Cliff Howard's rotary hoe and Albert Fuss's air seeder, Australian farmers have constantly and relentlessly improved their productivity.

It's an amazing history which has had implications for farming and food production all over the world: the rotary hoe alone transformed the lives of millions of people. And the innovative work is still continuing, often in specialised equipment such as broccoli-picking machinery or olive harvesters, any of which have worldwide markets.

There are many great stories to be told and, importantly, a tradition to be maintained. Visit the Gyral Implement factory in Toowoomba and there, out the back, is the original air seeder that Albert Fuss patented in 1956. You can see how he incorporated old bicycle parts (his

RIGHT River Murray fruitgrowers, who have a long tradition of 'gadgeteering', examine an early homemade apricot slicer.

LEFT Welder from the 1950s and still in use. The on-farm welder was a tool that started a million innovations.

ABOVE Albert Fuss's prototype air seeder. Note the bicycle parts …

children's, apparently) and other bits and pieces to make the prototype of an idea that would revolutionise crop planting worldwide.

It just sits there modestly, still in working order (of course).

In the late 1940s, in the aftermath of increased industrialisation of Australia during World War Two, on-farm arc welders became relatively commonplace. War surplus equipment started to flow to the farms and old generators from aircraft were used to generate the substantial current required to arc weld. Suddenly, jobs that would previously have needed a skilled blacksmith could be done simply, if not rather attractively. The ability to join almost any piece of iron to any other piece of iron drastically improved farmers' ability to adapt and rearrange their machinery, not

to mention using scrap iron for otherwise unimagined projects. The on-farm welder also brought with it the ability to reconstruct easily if you made a mistake. As one farmer said, 'We're not like other people who are frightened to modify things in case they destroy the magic.'

This started a culture of farm gadgets and improvement that has only recently started to taper off. Field Days and Agricultural Expos often show a dazzling range of clever ideas that exploit what is available to the ingenious farmer. Increasingly these farm gadgets are incorporating computerisation and sophisticated manufacturing techniques.

The farmers who appear in this section are typical of those who work in this brilliantly creative part of Australian life.

RIGHT Interior of the Murtoa stick shed in Victoria. Built as a wheat storage shed during World War Two, it was once one of the biggest buildings in Australia. It is no longer in use and will soon fall apart.

OMER'S LAB

'Just say what you need, I'll be there to make it,' says Omer. He's pleased to welcome people to 'Omer's Lab', an open-sided shed on Saroop's fruit block on the River Murray.

'I'm half a mechanic and Omer's a full mechanic. Between us we are more than a mechanic, we're one and a half mechanics!' says Saroop.

The Lab is your classic farm shed, strewn with old parts and tools, many of them stored in old dipping boxes (of the kind once used to dip apricots in preservatives) or old bread tins from the local bakery.

'Every year we have a clean-up,' says Saroop. 'All the junk around here that builds up, as we get a part off it, goes down to the back of the block. Then we gradually bring it all back up again as we need to find more parts.'

OMER'S LAB

A Sikh, originally from the Punjab in India, Saroop started out picking fruit, and by hard work over many years he gradually increased his land holding. He now has 400 acres, with 100 acres under citrus and 120 under grapes, all watered by a sophisticated computer-controlled drip system.

Saroop points out that like most agricultural businesses these days, in his line of work you have to get bigger or get out.

'I am concentrating on making the operations a lot more mechanical because labour is becoming expensive. For example, a spray unit that once only sprayed one side of a row of trees has been adapted to spray two sides. New machinery is also very expensive, and it's smarter to adapt an existing piece of equipment. That's

probably the most important thing we do in Omer's Lab.'

It requires a certain nerve to take the angle grinder to a piece of farming equipment worth $100,000. It's not unusual to find a million dollars worth of machinery in a farm shed. But growers like Saroop, never quite satisfied with the standard equipment, are always looking for ways to make improvements.

There's a long history to this approach. Back in the 19th century, local Agricultural Bureaux would meet regularly to check out one another's experiences, share ideas and visit successful or new agricultural ventures. More recently, annual Field Days and farm-gadget competitions have been powerful sources of inspiration not only for individual growers but for entire farming sectors. Modestly proud farmers

show off a particular piece of innovation or adaptation to their peers in the full knowledge that the idea will be copied and, better still, developed by neighbours and fellow growers. Off-the-shelf farm equipment is inspected and discussed, and probing questions asked.

This tradition of sharing ideas has been eroded in recent years as more specialised agricultural equipment has come on the market. Computers and digital controls are less easily 'fixed' with a welder.

'We modify things to make them work the way we want,' says Omer. 'When we need something, we make it. We want to save effort on the time-consuming things.'

'What Omer and I have learnt, we have learnt the hard way,' says Saroop. 'We don't want other people to have to go that hard way, so we are happy to share our experiences of how we modified that pruner to make it a five-way pruner.'

For farm-machinery manufacturers, innovators such as Saroop and Omer represent both a threat and a benefit. They are a threat because their probing questions might reveal enough for them to go back and build their own version, but they are also a potential source of new farming inventions.

Omer has been working for several years on his automated bud-cutting machine, a masterpiece of revolving cams and blades with a small motor cannibalised from somewhere. It has not been perfected yet but what's a lab for if you don't have a project on the go?

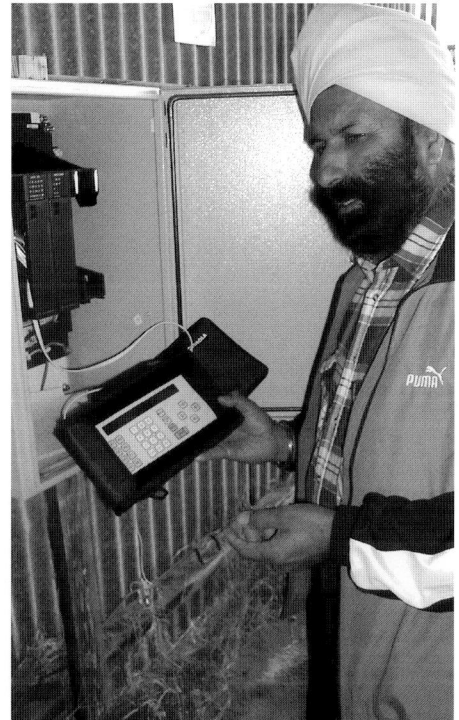

ABOVE Old dipping tins once used to treat dried apricots make perfect containers for old tools and other bits and pieces.
LEFT Saroop runs a state of the art crop watering system, which nicely complements an open-minded attitude to using technology.

SERIOUS HOME BREW

As I approach Cliff Kearney's property, a crop duster roars overhead, just clearing a couple of poplar trees. It's Cliff at work, preparing his paddocks for a crop of canola.

From his office — the cockpit of his Air Tractor — Cliff sees a lot of the country and as his work depends on the seasons he is keenly aware of the impact of climate change on the Australian landscape. That's one of the reasons he is closely involved in the production of biodiesel from canola oil.

Cliff, local farmer Ed Wilson and a couple of others joined forces to build an impressive transesterifaction plant. Mainly constructed from junk and other useful materials found on their properties, the plant can, given sufficient input, produce 1,000 litres of biodiesel a day.

THE FARM SHED
SERIOUS HOME BREW

'We thought: how hard can this biodiesel caper be?' says Cliff. 'So we looked up the internet, did some research and figured out the chemistry, talked to a few people and built this plant over a few weeks. A month after starting the project we were producing useable biodiesel that runs perfectly well in most of the farm diesels here — Landcruisers, bulldozers and harvesters.'

The plant uses old tanks, fire pumps, heaters and piping salvaged from all over the place. A storeroom full of new and used hydraulic equipment and an assortment of gauges and valves was invaluable.

Biodiesel is produced by a chemical reaction between vegetable or animal oil and an alcohol (usually methanol), often with caustic soda (sodium hydroxide) as a catalyst. This reaction produces a methyl ester (biodiesel), with glycerin as a by-product. At several stages, heating, cleaning and filtering are necessary.

'Our aim was to make on-farm production feasible partly because we were looking for some way of value-adding to canola crops. We want to be able to grow it, press it, process it and use the end product here on the farm.'

Cliff and his mates are not alone in their endeavours. A movement has sprung up across the country, converting waste vegetable oil into biodiesel. Virtually no fish-and-chip shop or fast-food joint throws out any old cooking oil any more: it is snapped up by home-brewers of biodiesel.

But there's a fly in the ointment. Customs and Excise want to tax their home brew to the tune of 38 cents per litre, arguing that it is a commercial fuel in the same class as normal diesel or petrol. Biodiesel makers are rightly annoyed at the intrusion of money grabbers into what they feel is a positive and beneficial contribution to the community.

While the economics and environmental credibility of biofuel remain hot topics of debate in Australia, with claims and counter claims frequently muddying the discussion, other countries, notably Brazil, are moving towards large-scale biofuel production. It makes sense in the context of soaring oil prices and world climate change. 'Australia is years behind countries like the USA.'

Cliff and his group have also built a small-scale plant which can make about 140 litres a day. They regularly take this smaller plant to Field Days to spread the good word.

From his early training in the workshop of one of Australia's biggest regional airlines, Cliff acquired the mechanical skills necessary to keep a broad range of farm and aircraft back-up equipment working efficiently. Aircraft mechanical engineering has a strong tradition of quality, safety and thoroughness, and a reputation for high-order problem solving. That tradition of resourcefulness mixed with the similar tradition of Australian farming serves Cliff well.

What is most encouraging about Cliff and his mates is the evidence they provide that an Australian culture of local resourcefulness is alive and well. What is discouraging is the knee-jerk reaction of various organisations to suppress such initiatives. Regardless of whether on-farm biodiesel production is viable from a strictly financial view (especially in the apparent absence of any planned large-scale production), Cliff and his mates should be encouraged because at least they're having a go!

RIGHT Cliff's transesterifaction plant — a classic DIY job with startling results.
BELOW Trial and error are everything in an endeavour such as on-farm biodiesel production.

LOOKING PAST THE PAINTWORK

If isolation and a shortage of skilled help are motivating forces behind the Australian tradition of resourceful problem solving, then Theo is a perfect case study.

In the early 1960s Theo purchased Thistle Island off the coast of South Australia. This fairly isolated 16-kilometre-long island had a human population of one — Theo, when he was there — and a few thousand sheep.

Farming a property of this size and this isolated is a challenge at the best of times. 'I enjoyed working on my own, but I was aware that a vehicle could roll over and trap you or something could go wrong. It made you a little careful,' says Theo.

Not only did it make for a certain caution but it made for some

LOOKING PAST THE PAINTWORK

interesting problem solving. Theo invented any number of small devices for opening gates or moving stock. He managed to devise a system for loading sheep onto a boat single-handedly.

His boat, the ketch *Hecla*, was famous in the region because of its unique bilge pump. Instead of the usual electric pump, Theo mounted a traditional windmill on the mast. It may not have suited the nautical design purists but a windmill, when there's wind, will pump water on a boat just as efficiently as it will on land. Old windmills are cheap and there are usually plenty around, and a boat faces into the wind while it's moored. It's a perfectly logical solution.

Another readily available item that Theo puts to good use is the rubber tyre. Just because tyre tubes are normally filled with air doesn't mean they can't be filled with water. That is just what Theo has done to create a pressure reduction chamber for water and, with slight modification, a pump. Where the rim would usually go, a diaphragm is inserted and made waterproof, effectively making it a small flexible-sided water tank which can adjust water pressure. With the addition of clacker valves the compressibility of the tyre turns it into a pump when the sides are pushed down. A tyre can be bent in half into a C shape, making it a large and powerful spring, which can be used to open a gate or move heavy objects.

All these solutions from a common object which until recently was thrown out as landfill or rubbish!

Another commonplace object ripe for adaptation and transformation is the truck differential. The crane in Theo's main shed is an elegant use of three old diffs that transfer high revving power from a one-horsepower motor to a low revving greater power, and also allow a change of direction of the crane. Diffs are used in other applications around the farm including a home-made power hacksaw, a cement mixer and a revolving welding vice.

Theo now lives mostly on the mainland but keeps up his interest in everyday mechanical challenges. His main shed is nothing short of spectacular. It's a whole landscape of machinery, some of it stretching back to his childhood, with the capacity of a medium-size engineering business. The 'things that will come in handy one day' are many, with paths worn through them like jungle trails.

The full tour of Theo's sheds takes about half a day. Even then you have to skim over a lot of important details and countless cleverly made devices, all of which have interesting stories attached to them. There's an aeroplane hangar with an immaculate Cessna 170 in it for travel out to the island, and several other sheds housing lovingly restored machinery including an 80-HP single-cylinder Crossley, a 1902 Blackstone, a 600-HP Mirrlees Blackstone and a 50-tonne railway-wheel lathe. Theo is currently restoring a very large 600-HP triple-expansion steam engine.

Theo has built some of the best playground equipment you'd see outside the Monash playground described earlier (pages 26–29), any of which would give a young person cause to think a little about how forces work.

Understanding how things work is the key to resourceful problem solving, according to Theo. 'You have to look past the paintwork. People see rusty machinery and all they can see is old junk. But if you can see why and how it was made, you can start to see new uses for it. At a clearing sale, everything has a value or a use. If you can't see the value then you aren't looking very hard.'

Theo's sheds are a testament to a productive life shrewdly led. He acknowledges that farming now requires a fair understanding of computer control systems for farm machinery. Use of the internet to plug into broader networks of information and markets is also useful.

'With computers and GPS, I can see the day when tractors will be working on their own and the tractor cab will be replaced by a remote office with operators moving from computer to monitor to phone, much like a captain on the bridge of a ship. That's what it will take to keep one-man farming viable.'

THE LAST JOBBING SHOP

When he turned 14, Henry's mind was already made up: 'All I wanted to do was go on the farm.'

So he left school to work on his father's farm, a mixed farming property in good steady country his great grandfather had settled in 1861. But his life was sent in an unexpected direction by fate in the form of a lathe Henry's father bought a few years later. Admittedly fate had some help: the purchase was made after a lot of pestering from Henry. It was the first lathe in the district and as Henry taught himself how to use it he gradually became known as a handy sort of fellow. He'd also got his hands on a welder and built a few things.

Farmers started bringing machinery around. 'A lot of their equipment was made in the US or Canada and we got a reputation for making

THE LAST JOBBING SHOP

parts that were better and more strongly made than the originals.'

Over the years, Henry bought more mechanic's equipment, and farming became part time. Eventually the mechanical work became full time.

Henry's business fills the time-honoured role once played by the local blacksmith: he's a useful community resource willing to repair anything.

'We're one of the last jobbing shops around. We repair original parts or construct heavier components. There's no money in doing one-off jobs compared to repetition work.'

Perhaps it's not so much the money as that he and his co-worker Mark, who trained as a fitter and turner, enjoy the challenge. No two jobs are ever quite the same. It helps that Henry has been a farmer

FOR ME, I KNOW THERE'S AN ANSWER BUT I DON'T KNOW WHAT IT IS YET.

and therefore understands the farming application of the machinery he works on.

The social context plays a part too. 'Customers are often friends and neighbours, so taking time to chat is an important part of business. There's a fairly humorous approach, so there's a bit of cheek backwards and forwards.

'Generally there are no secrets — if you don't tell them something, they'll find out from someone else.

'Our clients' expectations are high: they don't need equipment to break down when they're under the pump, such as at

seeding or harvesting. And we don't want comebacks either.

'Most major breakdowns we can fix in a day, although electronic and computer stuff is totally beyond us. We work on the mechanical side. If we fall back on a bandaid solution, we explain that's the case, and we fix it properly off-season.'

The dynamics between two people with slightly different takes on a problem can be interesting. Both Henry and Mark read a lot of technical manuals, but their differences in background show.

'In many cases, there's a beaten path you have to follow,' says Mark. 'At other times we come at the solution through different paths or different answers — but both can be right.

'Henry has some habits that would have got me a kick in the backside if I'd had them when I was an apprentice, but they are the sort of habits that can actually get you out of strife.

'We engineer in a fair safety factor, which sometimes makes the job a bit heavy, but at least it won't come back and bite us on the bum.

'A lot of ingenuity or nous or whatever you want to call it comes from taking an interest in the job or problem in the first place,' says Henry. 'You get some young bloke on the phone and if it's not a run-of-the-mill problem, they don't want to be bothered.' Henry is dumbfounded by this sort of attitude. 'You have to want to help. For me, I know there's an answer but I don't know what it is yet.'

RIGHT The obligatory shed dog, of course.

THE WOOLSHED

The woolshed is an Australian icon. It's the subject of innumerable folksongs and paintings, and a vast mythology of hard work, even harder drinking and other character-building activities.

The woolshed has been the birthplace of more Australian myths and legends than you can poke a stick at. It's a place where the hard men still compete to shear the most sheep a day, where a certain Australian male sense of stoic toughness was born. It was also the venue for the birth of the Australian labour movement in the great shearers' strike of the 1890s.

Shearing is still hard work — one of the hardest jobs in the world in terms of energy expended. And woolsheds are still no-frills places, with their own sense of humour, food traditions, allocated roles for the shearer, the rousabout, the classer, the cook and so on.

In the 19th and early 20th century, when Australia really did ride on the sheep's back and vast amounts of money were made by the pastoralists, there was stiff competition to build impressive woolsheds to flaunt that wealth. As a result, there are some huge and often unusual buildings in stone and corrugated iron to be found in the areas of Australia where the Merino was king — generally southern Queensland, New South Wales, Tasmania, western Victoria and outback South Australia.

RIGHT Ted has been shearing since he was ten years old and estimates he has shorn more than a million sheep. At 76 and after a broken back and quadruple bypass surgery, he can still shear 90 sheep a day.

As Peter points out in discussing his unique woolshed design (page 88), Australia is a land of horizontal elements, a sprawling, stretched-out, open place, and that's the quality possessed by many of the classic woolsheds. They are functional buildings and until recently have not been generally recognised as 'proper' architecture.

Whether they are ramshackle and patched together or well maintained, woolsheds have a certain style. Take Deargee woolshed (pictured). Although built in 1869, it looks not unlike a corrugated-iron version of the Starship Enterprise. Or take Old Errowanbang with its curiously shaped rooms, looking like a medieval fortress. There are hundreds of these buildings in many variations across the length and breadth of the land.

ABOVE Deargee woolshed, Northern NSW.
LEFT AND RIGHT Old Errowanbang woolshed, Central NSW.

In their way, they are our own corrugated cathedrals — the ultimate sheds, both in their built form and in the social history they carry.

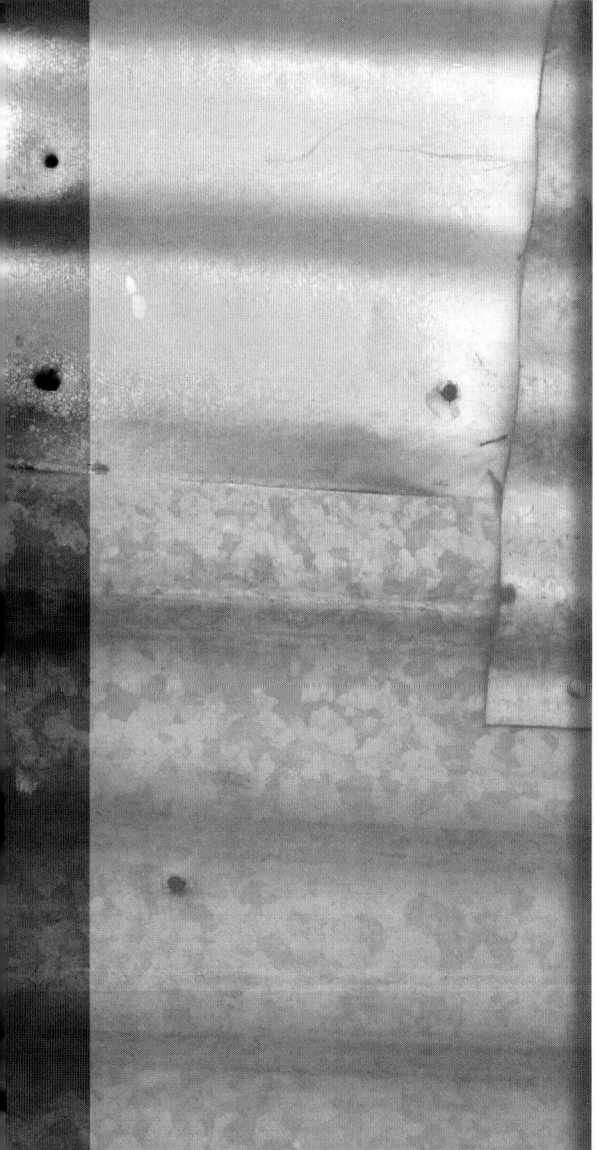

DEEPWATER

Hovering, almost shimmering, over a Riverina hillside is Deepwater woolshed. This dramatic building, which has won numerous awards for architect Peter Stutchbury, is a convergence of tradition, practicality, experience and patient planning that has become an instant classic.

When Michael Darling, the owner of the property, decided he needed a new woolshed he turned to Peter and presented him with a number of parameters, including a limited budget and an intention to create a more acceptable workplace.

With some experience on the land as a child, Peter had a fair grasp of how a woolshed worked but he also talked to many other people, including Andrew King, the manager (shown here in this photo). Peter camped on the site for four or five days before starting the design.

'We talked about what was wrong with the old shed, how to handle the wool bales efficiently, even the psychology of handling sheep,' says Andrew.

The result is a structure put together with a frugality appropriate to a working building yet with a calm balance and unmistakable elegance.

The structural strength of the Aramax deep-profile roof allows the eaves to extend unsupported some five metres out from the building, increasing the sheep-handling area and effectively providing a huge verandah. It also provides a large, uninterrupted internal space.

Nearly all the components are off the shelf — from the giant corrugations of the Aramax roof to the horizontal ventilation shutters (adapted from a commercial chook-shed design).

The entire building is bolted together — so it can be disassembled and recycled. Not only is this a building with impeccable environmental credentials, but those who work in it enjoy vastly improved conditions.

Old shearing sheds can be rough places to work: when it's 38 degrees outside, they can be more than ten degrees hotter inside. So passive cooling and heating was a very important consideration at Deepwater.

In hot weather, the entire building can be opened up and water (captured from the roof) allowed to percolate down expanded steel mesh to cool it, a bit like a giant Coolgardie safe. In cold weather, the body warmth of the sheep gathered under the shed is trapped and used to warm the shearing area.

It's even got a barbecue deck.

For Peter, designing a woolshed was an opportunity to reflect on the nature of the classic Australian building and its place in the landscape.

'Australia is a horizontal continent. Most other countries have strong vertical elements: we don't. The other factor that influences our buildings is light and shade. Our trees don't eliminate light in the way that dense European trees do. Our soils respond to light.

'I deliberately used white to reflect light but I also wanted to play with light and shade in a way that echoes the quality of the Aussie bush . . . light off water, light reflecting off country. So the "lightness" of the building is important. Heaviness doesn't work in Australia. It's about a kind of impermanence — human activity on top of an absolutely ancient landscape.'

Designing a shed presents some interesting challenges: Peter cautions

ABOVE Wide open spaces ... inside.
RIGHT The original shearing shed.

against thinking it's a simple task. 'A shed is a sort of universal building: it's effective architecture that says what it does. Traditionally the shed is an unrefined building which requires a certain acceptance. It may be unrefined in its finishes. But it is not unrefined in its intelligence.'

THE CORRUGATED
BROTHERHOOD

Chapter 3
THE CORRUGATED BROTHERHOOD

There is an idea in circulation that men who are shed struck are off being moody brutes: lonely, unhappy creatures who are up to something. They're not to be trusted.

And if the stereotypical man in a shed on his own isn't bad enough, men in a shed in a group are surely a recipe for disaster.

In truth, the reality is rarely like that, but it takes more than a flat denial to beat a stereotype.

The Men's Shed Movement has sprung up in the last few years and flourished to the degree that there are now several hundred groups around Australia. The groups vary widely in approach: some have arisen from a group of men sitting around deciding they could pool resources, give themselves a better social life and maybe do something constructive.

Others originate from a sort of honeypot approach, which can be a little misguided, as in healthcare or social services hoping to attract people they believe need treatment or care.

One aged-care company established a shed in one of their 'sunset facilities' and wondered why the men didn't flock to it. It turned out to be an empty shed with almost nothing in it — the woman behind the idea evidently thought men had some built-in attraction to corrugated iron and all they wanted was a metal structure. The idea that a shed was about doing useful things had to be explained to her. While the official shed stood there clean and empty, all the men in the place were down in the maintenance guy's shed watching him clean the lawnmower, ten or fifteen of them sitting around watching while one fella did the work.

Attempts to 'manage' men like this are a big worry. The people they are hoping to attract are likely to feel patronised or 'worked on'. In one Vietnam vet's words, 'I felt like some kind of laboratory rat.' People have good bullshit antennas.

A retired but active cabinet maker told me he went down to the local community shed hoping to lend a hand and was told by a patronising social worker that he would be allowed to make a set of pull-along wooden toy trains in the first three months. That's quite an insult to a man with 50 years' cabinetmaking under his belt.

The young social worker didn't have a clue about cabinetmaking or any other complex manual trade, but he was on to the fact that the social-shed caper was a good career move. A few years ago it was all tree hugging and rolling in the mud; now it's stuff in sheds. Perhaps the worst insult comes when a ham-fisted system is brought in to check whether these old men might be dirty old paedophiles. A number of men are so appalled at being suspected of such crimes that they leave and never return,

which in turn feeds the suspicions of those who enforce such systems of control.

Some community shed organisations, on the other hand, realise that there is a vast reservoir of skilled labour waiting to be tapped among retired tradesmen. At its best the local Men's Shed becomes a sort of reinvented Mechanic's Institute where knowledge is freely given and shared: there is huge potential there.

The Nambucca Valley Men's Shed also provides an interesting way of helping young men grow up: sometimes there is just nowhere for young men and boys to find direct basic examples of how to be a man. There are plenty of dodgy examples in the media, and schools aren't in the business of manhood creation. The face-to-face, no-bullshit learning of what is and isn't acceptable happens very well in the context of the Nambucca Valley Men's Shed and hundreds of others around the country.

Most older men are keen to help young men and boys grow up into people who have purpose without greed, self-worth without arrogance, and who are just plain decent and civil to other people. And most young men are looking for that chance, even if they sometimes don't know it.

The other unspoken need that the community shed fulfils is the big L: loneliness. It's not easy for men to admit to isolation and loneliness. Television may have cosy afternoon chat shows that endlessly dissect the finer (sometimes weirder) points of human relationships but this is hardly riveting to most men. Men who have led a busy working life are often more comfortable having a task to achieve and get a big kick out of

being part of a team, gang, clan, crew, tribe, group, company — call it what you will — that successfully undertakes that physical task. The desired benefits of social interaction, such as talking through problems, tend to follow naturally from those activities but it would be a mistake to underestimate the fundamental importance of the job being undertaken. It needs to be genuine and complex and allow people to put something of themselves into it.

Given that work, in the last few decades, has broadly become less physical and more about social skills and pressing small buttons of various kinds, it's not surprising that many men feel a bit lost and isolated. And there are becoming fewer of these team activities for men to fit into relatively easily.

It's no wonder that community-shed funding has become an election issue in a couple of recent state elections. Politicians, who can usually sniff a change in the social fabric faster than a dog can smell three-week-old pee, have become aware that there is a yawning human hole that they have not been previously aware of, and it may need filling. In addition, those with a professional interest in the community shed have become increasingly organised by holding conferences and exchanging information on the internet.

The social shed will be an interesting place in the future as more places like the Williamstown Shed Club or the Nambucca Valley Men's Shed spring up. They will change our social fabric. Watch this space. www.mensshed.org

ANY PORT IN A SHED

Like many brilliant ideas, the Williamstown Shed Club had its beginnings at a dinner party after a few too many red wines. Someone said that while the women of the area knew each other well through schools and children and so on and men knew each other to say hello, the men didn't know each other well and didn't have a forum to develop their mateships further.

It was decided that the humble shed would be the best place to develop these local bonds — and the Williamstown Shed Club was born. Members would host a meeting every six to eight weeks in local sheds. Rather than simply gathering in a shed and drinking till they fell over, the founding members decided to have a small amount of formality. Apart from providing a convivial atmosphere and a bit of food, the host must prepare three sessions: a game, a tool session and a men's issue.

ANY PORT IN A SHED

A game. As men are all warriors (or so we think), the game must allow for the combatant in us all to come out. It can be anything: table tennis, darts, a quiz, backyard golf. Whatever the game, a winner is to be found and honoured.

A tool session. The host must display and discuss his tools and tell stories about them. He may discuss his oldest tool or his favourite tool, for example. Of course, the main event of this session is the power-tool count where a nominated member must count all legitimate power tools in the shed (the highest is currently 14).

A men's issue. Here the host must lead discussion on issues facing men — anything from work stress and back pain to premature ejaculation and penis size anxiety is fair game.

Throughout all of the sessions heckling is encouraged to make sure nobody gets too serious. While there are some serious questions and answers, a fair bit of bullshit is flung about. There are your obligatory ribald jokes. A barbecue is thrown in for good measure, and drinking does take place, but only to enable the blokes to maintain the pace for the night. It's all very blokey, but that's who's there — blokes.

The Shed Club manages itself very well, keeping organisation to a minimum yet not letting things fall apart. It has guidelines rather than rules. One of those guidelines is that members or visitors must not drive home. Accordingly at most meetings a good collection of pushbikes clutters the front of the shed.

Since the club's inception in 2002 membership has climbed to around 20 card-carrying sheddies. Given the conditions of membership that's no mean feat. Most meetings herald a hard core of 10 to 15 members and visitors, and it's usually the same four or five hardcore members propping up the port bottle at two in the morning before wobbling off into the dark on their pushies.

Once a year there is an Awards Night when members receive coveted Sheddies — framed certificates to display proudly (in the shed of course): Best Shed, Best Men's Issue, Best Game, and of course the Gold Sheddie awarded for the Best Meeting of the Year.

Similar shed clubs have been formed in other parts of Australia, and there are even the beginnings of an international movement: clubs have started up in New Zealand, England and Germany. It seems that there is a need all over the world for the kind of camaraderie that is born in the shed.

What it comes down to is that the Williamstown Shed Club helps its members feel a part of a village-sized community. It's like being part of a tribe — not something that most people living in suburbs get to feel. For that reason, a club like the Williamstown Shed Club is one of the best uses of the Australian backyard shed you're likely to see.

As Mike, the self-appointed PR manager says, the spirit of the Williamstown Shed Club is summed up in its slogan. 'Any port in a shed.'

To form your own shed club or find out more about the Williamstown Shed Club, check out the website at:
www.shedclub.com

IT COMES DOWN TO RESPECT

Unlike many other community men's sheds, the Nambucca Valley Men's Shed is set in an industrial estate. According to Stu, the Shed's project officer, that setting has some benefits. 'It's not just for the knowledge and skills around us, but it shows the young blokes that this is part of the world of work and is a serious undertaking. The older blokes like it for similar reasons — they are involved in an activity worthy of respect.'

The Shed undertakes a wide range of projects such as boat building, housing components and anything that their limited resources but unlimited enthusiasm will allow. Like many similar sheds, the tangible things being made are only part of the story.

'We're a community focal point,' says Stu. Some people find their

IT COMES DOWN TO RESPECT

own way to the Shed and others are part of bigger schemes — community work placements, disability programs, probation and parole obligations, school-to-work programs and so on.

'We live in a big retirement area, where older couples move to a retirement village and don't have the social networks they once had. There are reasonably good social supports for women but for older blokes it's not so easy. For men it's usually been the pub or the club — which not everyone wants to be part of, especially if they don't drink much.

'We've got an open-door policy here. Apart from myself, no one needs to know how participants come to the Shed, who sent them or why they are there until they're happy to divulge that themselves.

'Everyone at the Shed is a participant, some are volunteer participants, others client participants and some are both. It's about everybody being given the same opportunities without preconceived labels or stigma.'

Easygoing fellowship and a practical work ethic are carefully mixed and monitored by Stu and the other more experienced men.

'Young fellas in particular need to know that there is a difference between school and work. You're showing them that fulfilling activities are possible and open to them but there's also the reality of factory and workplace life. Sometimes you need to knock a few corners off, set a few boundaries. But it's no different than what existed naturally in many workplaces years ago.

'It's a bit like being part of an extended family with a few pseudo-grandpas. There's sometimes intergenerational mistrust at first — "What could this old codger possibly know?" or "Why does that stupid kid listen to that horrible music?" But after a while you cut through the crap.'

Many of the kids who come to the Shed are from single-parent families with no experience of using tools or making things, or any real idea of what is acceptable behaviour among men.

'We're showing them what blokey behaviour is all about. Everyone's equal here until they stuff up. Then we tell them, "You'll be treated as an adult but if you make a mistake, tell people. Aside from basic rules of commonsense, it's a respect thing. Respect for yourself, respect for each other."'

The older men appreciate the feeling of 'handing on the baton' to the next generation and it makes them feel good about what they've done.

'Initially people suspected us of misogyny, that we were running some sort of anti-women thing down here, but the results we've achieved have swung opinions around. Some mothers of young men have been ecstatic over the way we've changed their sons' lives.

'Not that it's all been plain sailing. There are many angry men out there and violence has to be dealt with carefully. Sometimes the frustrations of what's going on in the blokes' lives come to the surface. Having a supportive group of men who are non-judgemental, can empathise and give sound advice helps most to realise that acting out and becoming violent is not going to serve anybody's best interests . . . let alone the

THE RESULT IS THAT WE'RE BUILDING BETTER BLOKES AS WELL AS BUILDING THINGS. AND THAT CAN ONLY BE GOOD.

individual who's already suffering within himself.

'The problem we've got for men's health issues in Australia at the moment, is that we've parked the ambulance at the bottom of a cliff — the services are there to pick up the pieces once the bloke goes over the edge. And when they do go over the edge, they often take their partner, their wives, their kids, friends and family with them, metaphorically speaking. All that we are trying to do is move the ambulance from

the bottom to the top of the cliff.'

Stu is a passionate advocate of the local shed and he's not alone. When the Shed's very modest funding (most of these institutions run on tiny budgets) was not going to be renewed, they looked around for support — and found it in a big way. Some 3,500 locals signed petitions and well-known singer John Williamson offered substantial backing.

For Stu, it was a realisation that the low-profile work they have been doing since 2000 has been acknowledged.

'When we started we realised it was important that the community pointed us in the direction they wanted to go, that we do what they wanted us to do. The result is that we're building better blokes as well as building things. And that can only be good.'

THE MAGIC OF FLIGHT

The magic of heavier-than-air man-made flight is to a special obsession. It's hard to communicate the joy of making a fragile craft out of balsa wood and tissue or cloth, and sending it to float gracefully in the air. There is something inexplicably pleasing about seeing a man-made aeroplane, even if only a model, take flight for the first time. It's as if a part of the spirit of the model's maker has lifted off into the ether too. Ivor should know about these things. The Doonside Aeromodellers Club has been meeting in his shed for some 60 years. Ivor feels that he has put in a decent effort and at the recent meeting (shown here) he officially stopped taking the main responsibility for the club.

This shed is a wonderful space, packed with small-scale history: dozens of models and prototypes hang from the ceiling, and the walls

THE MAGIC OF FLIGHT

WE DIDN'T HAVE A LOT OF MONEY AND WE MADE ALMOST ALL THE PARTS WE NEEDED FROM SCRATCH — WE HAND CARVED PROPELLERS FROM BALSA.

are lined with tools and aeromodelling magazines going back to the early days. Ivor's shed is virtually a sacred site dedicated to aeromodelling.

Ivor's passion was ignited in the 1930s when the *Women's Weekly* published coupons for a model aircraft kit designed by Norman J. Lyons, a keen early aeromodeller. The magazine

expected to sell 2,000 of the kits. They sold 25,000.

'At the time Australia was probably the most aeronautically conscious nation on earth,' says Ivor. Charles Kingsford Smith, Charles Ulm and Bert Hinkler were world-renowned Australian aviators and with the birth of organisations such as QANTAS and the Flying Doctor flying seemed very modern and glamorous. Model aeroplanes shared some of that glamour, yet were reasonably affordable. When Ivor got the components for his model, he was hooked.

'Even when the rubber that drove the propeller wore out and snapped, I still climbed up to the highest point of the roof and flew the plane like a glider.'

After the war (which, from Ivor's perspective, is an interesting story in

itself), he maintained his interest in aeromodelling.

'Three of us started Doonside Aeromodellers Club. We didn't have a lot of money and we made almost all the parts we needed from scratch — we hand carved, propellers from balsa.'

Ivor estimates he has carved 25,000 propellers ('at a penny per inch') over the years.

The club attracted kids like flies. 'We were open twice a week for the first 40 years and in that time I suppose a thousand kids would have come here and learnt something about building models. The purpose of the exercise wasn't to make them lifetime aeromodellers — although a few dozen of them are. No, what we wanted was to provide a place to come and learn for as long as they wanted. Some of them are still coming after

50 years. Some of those who came here as kids have sent their kids and some of the original kids are in their 60s now.'

It's an extraordinary legacy to leave: not that Ivor is stopping any time soon. He plans to keep running classes at local schools on how to build small flying models. Thanks to his long experience, he is able to keep costs to a very modest level. The thrill of seeing an aeroplane fly for the first time is still as big a kick as it was 60 years ago.

'I just hope I haven't sent any kid backwards.'

RIGHT After 60 years Ivor's shed is still a regular scene of busy exchange of information, trading materials and parts and shared passion for flight.

AN IDEA WITH TRACTION

Every now and then a shed project hits gold. It might lead to a brilliant invention or a big technical breakthrough, or it might be golden because of the way it affects people's lives. The Sunrise Surfers Wheelchair Trust Rotary project hits that last kind of gold.

Des, Daryl and Keith are behind the brilliantly simple idea of converting old bicycles into simple go-anywhere wheelchairs that won't break down. Their wheelchair is a plastic chair on a platform with a bike wheel on either side. The idea of a wheelchair has been stripped back to its design basics: it's not a high-tech solution, nor does it have to be.

At the time of taking this photo, the Trust was about to deliver its 4,000th wheelchair.

The idea is down to Des who, while on holiday in Fiji, was shocked

AN IDEA WITH TRACTION

to see crippled kids with no way of getting about. 'I thought to myself that if you could give them a wheelchair, then their lives would get a bit better . . . I thought about it for a while, and the idea came to me one night: use old bike frames.'

Des, a cabinet maker who works in building and shop fit-outs, spent the next 18 months refining the idea, with the main criterion that his wheelchair had to be as cheap as possible. The only significant design change in the nearly ten years since the idea was hatched has been to add a castor wheel in the front. The tyres are filled not with air but with high-density foam (generally known as 'swimming pool spaghetti'), enabling them to ride over bad roads and thorns without the risk of puncture, which would be impractical to repair for people with minimal resources.

I THOUGHT TO MYSELF THAT IF YOU COULD GIVE THEM A WHEELCHAIR, THEN THEIR LIVES WOULD GET A BIT BETTER . . .

Nor is there a shortage of bike frames for little or no cost, as these days old bikes are frequently thrown out by people who don't know a good thing when they see it.

The 'Africanised' go-anywhere, unbreakable wheelchairs were well received and hugely appreciated. A map of the world on the shed wall shows an impressive range of countries where they have ended up.

To get the idea up to this scale of operation Des and his colleagues in the

local Rotary Club first secured a double garage, then another, then three. They now have a 360-square-metre shed salvaged and rebuilt from the local garbage-truck repair depot.

Along the way they have received excellent in-kind support from local businesses.

The actual fabrication of the bikes happens at several locations around Australia (thanks to Toll Transport) but they are centralised for distribution in the Gold Coast shed.

As community-based sheds go, this one plugs into the idea that men want to be involved in something that has a bit of traction — something that makes a tangible difference to people's lives.

'It's evolved into a huge thing for retired men around this area,' says

Des. 'Here in the Gold Coast there's nothing to do if you're an active retired bloke. By being involved in this project, we're giving them back a life. We've got one fella who's 87 years old and has built hundreds of wheelchairs. Family members tell us that involvement in the wheelchair project brings about some really positive changes in these blokes' attitudes: they're over the moon after a day down here.'

Distribution of the wheelchairs overseas presents a whole different set of challenges.

'We can fit exactly 274 dismantled wheelchairs into a 40-foot container . . . but at the other end we have to be careful about import rulings and customs and so on. We've overcome most of those problems one way or another.'

After the 2004 tsunami disaster, Des gave some thought to applying the same sort of innovative resourcefulness to building lightweight housing for disaster areas. Already he's had some success with a small house in Vanuatu and a schoolhouse/community centre in the Phuket area of Thailand for 43 tsunami orphans.

'It comes down to necessity. I could see the need for the wheelchairs, low-cost housing and now the low-cost school buildings. I start to work on a design and slowly it evolves and comes to light . . . with, I guess, some help from above.'

RIGHT This list of countries with their names casually crossed off shows the extraordinary range of the club's activities.

A SHED TO BUILD A COMMUNITY

There is no such thing as an average shed: perhaps that is their attraction. If there is anything that defines the Australian shed, it's that a shed is a functional temporary building.

For Mariano, a leader in his Sudanese community, the backyard shed has a function quite different to the more well-established migrant communities.

'Our shed is used mainly to store items such as clothing and cooking utensils for newly-arrived refugees,' says Mariano. 'When people turn up, they often need things immediately. So they come here and have a look through the shed to see what they can use. It builds their confidence that someone is able to help them like this, which is sometimes greatly needed as many refugees come from traumatic backgrounds.'

A SHED TO BUILD A COMMUNITY

Mariano's backyard is used for many barbecues and gatherings, another contribution to the long process of consolidating a community and putting down roots in a new country.

'I was born in Sudan in a typical tribal society. For me, it was a difficult decision to leave my people, to leave the land in which I had developed affinities since childhood. That's never an easy thing . . . but after a while you realise that in a way you're lucky because you have come to a society at peace.

'I think for people like us, we don't build a shed in order to keep cars or to work in there as most Australian blokes do. We use our sheds in order to build a community.'

Photo: Aaron Gully

ABOVE New arrivals making use of the resources stored in the shed.

SETTING UP A NEIGHBOURHOOD MEN'S SHED

Like to try setting up a social or community shed in your neighbourhood but not sure how to go about it?

First of all, decide how public you want to be. You may prefer to set up a simple informal shed similar to the Williamstown Shed Club. Many shed groups such as this are often started by putting up a notice in the local hardware store. Woodworkers, in particular, have an excellent network operating on the internet — and are a sociable bunch. The Woodworkers Forum (www.woodworkersforum.ubeaut.com.au) is a useful, and most amusing, site at which to start.

Alternatively, you may want to set up a shed that will have a higher profile in your community and possibly make use of some otherwise rarely used building at the back of a school, church or an old council property. Start by determining who actually needs a shed. Discuss it with your local council representatives. They may have a policy which acknowledges that retired men, men with disabilities, unemployed men and other groups have needs and they may have funds to help. Similarly, state and federal government agencies such as the Department of Veterans' Affairs or non-government organisations such as church organisations and charities may have some funding at their disposal (or be able to seek funding on your behalf). Local Service clubs have been particularly effective and busy for many years in this field.

In seeking assistance, it's useful to point out that a good local shed can deliver financial benefits to the broader community. Such benefits might range from keeping kids out of jail, to keeping old fellas out of the health system, to identifying local skills not being used that could be of benefit. Perhaps there are older men skilled in car repairs or engineering who could offer community assistance. Most people involved in these shed projects get a big kick out of them, regardless of how or why they are participating.

Before setting up, also gain advice on issues such as occupational health and safety, workers' compensation, insurance, liability and so on.

Lane Cove Community Men's Shed (check out their website at www.menshed.org.au) and Mary McKillop Outreach (at 1B Thomas Street, Lewisham NSW 2049) are both able to give good advice on how to set up a community shed.

THE CAR SHED

Chapter 4
THE CAR SHED

Some of the largest and most complicated sheds in Australia are to be found among the lovers of two and four wheels. When you start building a separate trophy room onto the clubroom, or are thinking about installing a second hoist, it's getting serious.

For many men, fiddling around with the car is the form of technology which they learn most about and on which they are most likely to experiment.

The car is a separate world, a thing with its own integrity and personality.

It is still an amazing thing that when a car comes to the end of a factory production line and the key is turned in the ignition, what was just a collection of panels, lumps of metal, plastic and glass, bolts and bits and pieces will spring into life. There's a certain attraction and pleasing logic in dealing with a closed system like a car: you can see how one part influences or controls another or how the whole sequence works. You can see how cleverly some engineer has constructed a system — or in some cases how dumbly. You can focus on a car, concentrate on it, work on it and it will usually reward you.

And if you've done well, in the end it will go like the clappers, anywhere you want to point it. A car is an escape clause from the rest of life.

CAR + SHED = GARAGE

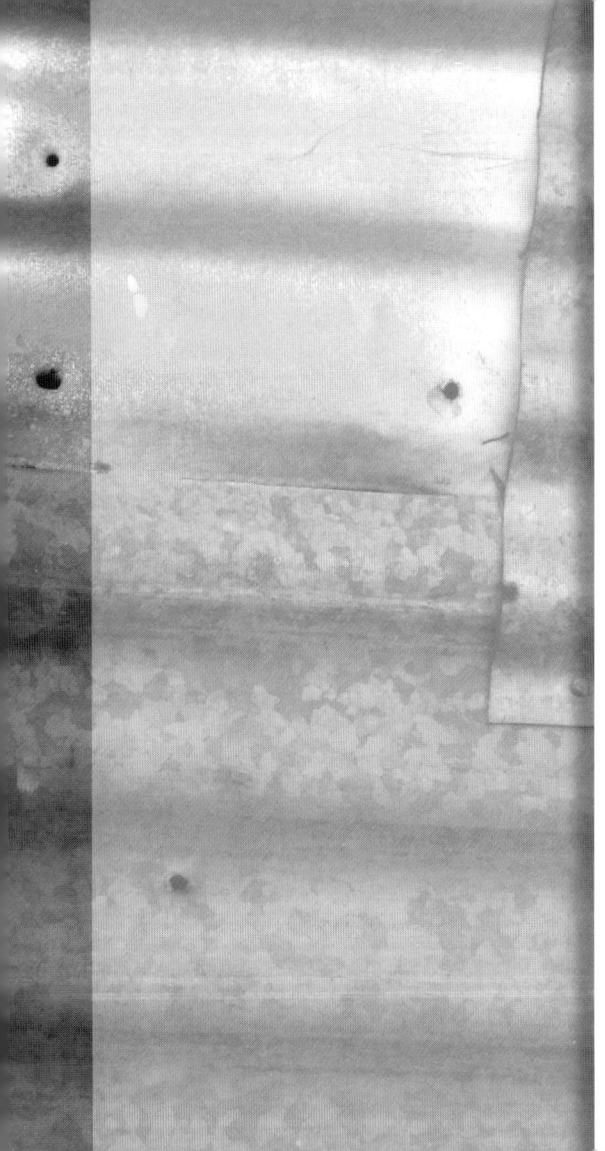

THE MODEL A KINGS

Bill and Lindsay run a 90-acre beef and dairy property in Western Victoria, where a drop of rain is not uncommon. They needed a farm vehicle with a bit of traction in a muddy paddock, so about 55 years ago they joined the modified back half of a Bren Gun Carrier to the front end, engine and chassis of a Model A Ford from the late 1920s.

The result is a very practical vehicle which has a split drive similar to the power take-off on modern conventional tractors. By carrying a generator on its back, Bill and Lindsay can have access to power out in the paddock. This means they can weld things, saw wood and get light at night a long way from the house or shed. Half a century after it was built the vehicle is still working fine. As Bill says, it's better than having the horses do all the work.

THE MODEL A KINGS

Their main car is also a Model A Ford, which their father bought new in 1930 for 210 pounds cash. They've still got the receipt.

The Ford has been around the clock a few times and there have been minor cosmetic alterations such as bare running boards — the calves chewed the rubber off — but other than that it's in remarkably good nick for a car that has been in frequent use for 75 years. The brothers have replaced the clutch a couple of times and keep a spare Ford motor on hand just in case.

In the shed is a large industrial lathe which they use to manufacture car and farming-machinery parts as needed. Between the lathe and the welder, most things that are used around the farm can be made or repaired using cannibalised

parts: they don't buy new equipment. The fire trailer, various loaders, scoops and minor items of farm equipment are all ingeniously made from existing parts, which the brothers are pleased to show off in a modest fashion. Frugal farming is an unbroken tradition on their property.

They've never been formally trained. 'Welding? Learnt that by trial and error. And I took the skin off on the lathe a few times too . . .' says Bill. 'You just learn things as you go. You have to have a go at least. But television — that beat me. Too complicated.'

Bill and Lindsay still milk the cows regularly as they have done for over 60 years. They tried spuds and pigs for a while but prefer the routine of caring for the cows. It's a quiet life.

ABOVE The brothers' approach to adaptation doesn't stop at cars. Their farm is full of mechanical innovations such as this delver for making drains and cleaning ditches.
RIGHT Henry Ford would be proud of the way Bill and Lindsay have looked after his product.

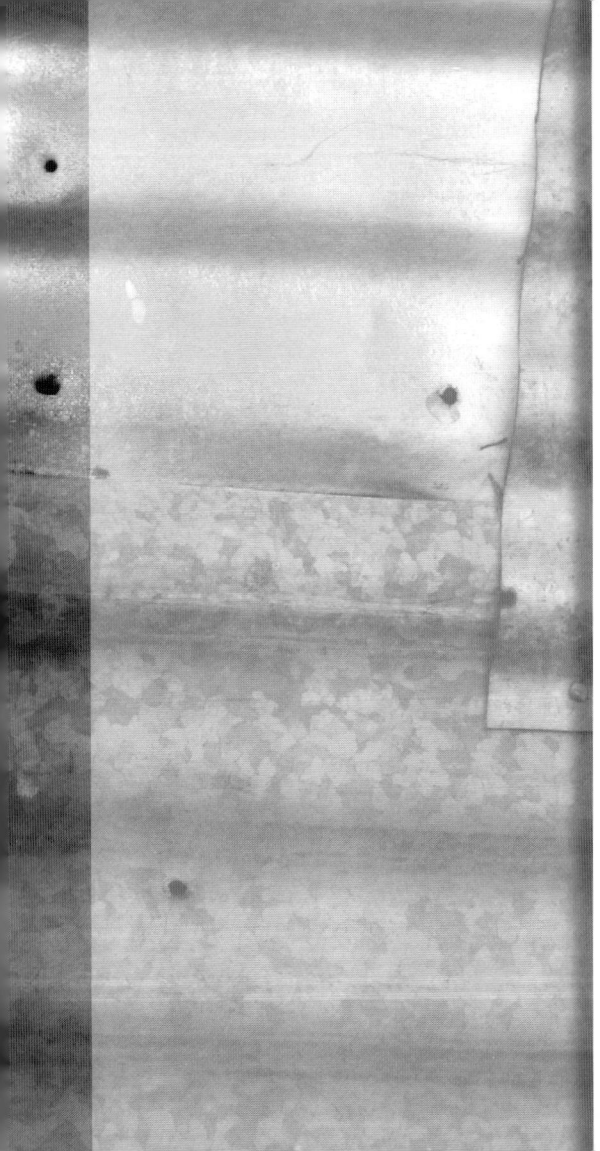

THE ULTIMATE

To call Mark's workshop/showroom/games and trophy room and club a shed is probably stretching it a bit. It takes up most of his backyard, a full 26 squares of ultimate automotive pleasure palace. By contrast, his house is a mere nine squares.

He's been showing cars since 1986 and his packed trophy room attests to his considerable success.

Although he's an electrician by trade he has an engineering background and does all the work on his cars except the paintwork and upholstery. For that purpose his shed has a fairly complete machine room that includes lathes and milling machines and of course a full size hoist.

Closest to the camera is a 1969 Camaro which was rebuilt using Holden HQ parts.

'I took it to all the shows and it won everything . . .' Mark says, almost dismissively.

In the middle is a real rarity and a pinnacle among hot rods: a Boydster 2 designed by Boyd Coddington, generally considered the world's leading designer of hot rods. Based on a 1932 Ford Roadster and one of only two in Australia, it represents nearly three years' work and a large amount of money for Mark. It has a 101-block 350 Chevy engine connected to a Turbo–400 gearbox, delivering a massive 300 kW of power. It has to be one of the deadliest hotrods in Australia, and has won its share of trophies for Mark.

Furthest away from the camera is a 1932 Ford 5 Window Coupe which won Grand Champion at Summernats in Canberra in 2003.

For Mark, embarking on a car-restoration project is an all-encompassing creative task that can dominate his life for years on end. It requires meticulous attention to detail, a great deal of technical problem solving and a strong sense of good design and balance. The result is a car of outstanding beauty and immense power.

'After three years' work, it becomes very emotional: you put a lot into the car. You've worked on every single millimetre of it. When I was working on it, I couldn't wait until it was finished. Now it's done, I've got nothing to do . . . and I can't sit and watch television.' So no doubt Mark will find another car to work on or a similar challenge. Meanwhile, he can lie back briefly in the trophy-stuffed games room and dream about the next big thing . . .

AFTER THREE YEARS' WORK, IT BECOMES VERY EMOTIONAL — YOU PUT A LOT INTO THE CAR. YOU'VE WORKED ON EVERY SINGLE MILLIMETRE OF IT.

LEFT Every detail of Mark's cars is not only superbly finished but shows a keen eye for classic car design.

ABOVE If you are going to have a trophy room in your shed, you may as well fill it up to the brim. Mark's cars have won an astonishing number of trophies.

CARS ARE MY LIFE

'Cars are my life. Everyone knows that,' says Phu, shown here with his brother Phuong in their father's shed. Phu's Nissan Skyline and Phuong's Subaru WRX are part of the Japanese performance-car culture that has grown up in recent years.

The Japanese performance car has spawned a group of car enthusiasts. Japanese sports cars underwent a rapid development in the 1980s and '90s to become highly competitive yet still reasonably affordable. Their competitiveness — at times matching or beating European sports cars such as Porsches — gave them a cult status, especially as they could only be imported in small numbers at first.

'They are the best value-for-money performance cars around. They've got everything electrical you could put into a car and they are easily

CARS ARE MY LIFE

worked to make an affordable go-quick car. They don't need big mods to perform well,' says Phu.

To Phu's way of thinking, his cars are in better than factory condition. They are immaculately looked after and many of the standard factory parts have been replaced with higher performance equipment. It's not unusual for car owners like Phu to spend up to twice the original cost of a car on modifications. The audio gear on this one has sufficient bottom end to push a breeze in your face with every bass note.

Phu can listen to this deadly sound system while he spends much of his Saturday cleaning and polishing the car. A range of 'car cosmetics' along the wall plays a part in Phu's weekend ritual. The end result is a low, gleaming spaceship of a vehicle: to step into the car is almost a science-fiction experience. To sit in the muted cockpit is to be surrounded by quietly flickering blue LEDs and dials, ready to blast off as the finely tuned muffler throbs deeply in idle. It's out of this world.

According to Phu, Sundays are for cruising. When you spend this much time and money on your car, there's no point in having anything mechanically inferior about it. This makes it a bit tiresome to be pulled over and checked out by the boysin blue. This happens to Phu quite often, and it costs $66 every time to have the car cleared by the government registration authority. This unwanted attention must take a lot of restraint on the car owner's part.

Phu and Phuong both learnt a lot about cars from their father who, according to them, had the best looking Datsun 200B around. When their father got rid of the old 200B, his sons persuaded him to buy a Nissan Silvia 1.8 Turbo. He still does the occasional accidental 360, according to the boys.

ABOVE There's no mistaking whose car this is.

TOP RIGHT The cockpit of Phu's car is a science-fiction experience.

BOTTOM RIGHT Car cosmetics.

ALL RULES SHOULD HAVE EXCEPTIONS

Some of the very best sheds and workshops around the country belong to vintage-motorbike owners and restorers. There's something about the scale of working on an old motorbike: it's roughly the same size you are. Bikes somehow easily acquire a sort of personality.

An old motorbike, especially if it dates from around World War One, is an open-book lesson in how technology develops. From the way that these old motorbikes are built, you can see that someone thought of putting a small internal combustion motor onto a pushbike frame — and just took it from there.

Colin and Catrina — a father–daughter team — have a workshop that includes a number of beautiful pre-World-War-One bikes and a range of others including 1920s AJSs and Colin's grandfather's Indian.

ALL RULES SHOULD HAVE EXCEPTIONS

The Indian was given to Colin by his grandmother when he was 15, when it hadn't been running for many years. In the 1970s he rebuilt it and got it on the road again.

There's no doubt that it's pretty unusual to see a father–daughter team working closely in the shed.

'What happens is that I do a lot of the spanner work and try to work out what we need to do, and Catrina makes what we need on the lathe,' says Colin.

Catrina's enthusiasm is genuine: she studied the practical parts of a Bachelor of Engineering course at TAFE for a couple of nights a week over a two-year period.

'Grandpa was doing a lot of this work when I was younger,' says Catrina, 'and it didn't look too hard. I thought: "I can do that."'

GRANDPA WAS DOING A LOT OF THIS WORK WHEN I WAS YOUNGER, AND IT DIDN'T LOOK TOO HARD. I THOUGHT: 'I CAN DO THAT.'

'I ended up liking it a lot more than I thought. We spend a lot of time refabricating parts. I particularly like seeing how different manufacturers came up with different solutions. It's still a good challenge and it's very pleasing when it all works.'

While their collection is clearly valuable, both Colin and Catrina are keen to point out that it's been created without spending huge amounts of money.

'Everything you see here we make or we get through the "friends economy". We make them, paint them and get them running well and reliably.'

Catrina is also a keen bike rider: 'When I was studying and getting stressed and anxious I'd love to go for a 30-kilometre ride, stop somewhere and eat some Country Women's Association soup.'

Her favourite bike is her 1926 two-speed Scott.

Colin, without saying much, is clearly pleased to have a good team. It also helps in the problem-solving aspect of their work.

'The fact that there are two of us is good because we think differently. I'm sure I visualise the finished job quite differently to her. It can be very interesting how two minds come at the same problem.'

WE LOVE GALVO

Like the warbling of magpies in the morning or the screeching of galahs at dusk, the sound of rain on a corrugated-iron roof is part of Australia's audio landscape.

It may come as a surprise to some people that elsewhere in the world corrugated iron is often regarded as a crude and even vulgar building material. In Australia it has somehow come to have a curious poetic quality: it is a welcome drum when the rains come, but it is also a light, strong skin that rusts and buckles but always endures — until it finally gives up and its flakes slip easily back into the rust red landscape from which it came. It fits perfectly into a corrugated country. It also features in the occasional tragedy such as when bushfire whitens and buckles it like dried leaves or a cyclone wraps it around trees like grass after a flood.

The history of corrugated iron is not a very long one.

Henry Robinson Palmer (1795–1844), a London civil engineer and founder of the Institute of Civil Engineers, is credited with having first used corrugated iron as a building material in 1821, although the corrugation of iron to strengthen it was already an established practice.

Corrugated iron was soon found to be an excellent material for use in utilitarian structures such as factories and railway stations. Curving or bending the sheets was found to strengthen it and made for more elegant structures. The drawback of its tendency to rust was partly overcome when in 1837 a Frenchman named Sorel patented a process for dipping iron into molten zinc. Sorel named this process 'galvanising' after the Italian scientist Luigi Galvani.

FAR RIGHT One of Australia's oldest corrugated-iron buildings, in South Melbourne.

One of the most successful early uses for corrugated iron was in the production of prefabricated houses and buildings for the American, African and Australian markets. These prefabricated buildings started appearing in Australia at about the time of the gold rush of the 1850s.

One of these early buildings, manufactured in Scotland and dating from 1853, is still to be seen at 399 Coventry Street, South Melbourne. This house is now owned by the National Trust and is in good condition. It was made with a thick-gauge iron and the corrugations are very large by today's standards.

Corrugated iron rapidly established itself as an effective construction material for use in homes, schools, churches, hospitals and factories when and where stone, brick or timber was not feasible. It was carried ten or

20 sheets at a time on camels in the outback. Only in the last few years have people started to appreciate the history of this early corrugated iron and collect it.

In 1921 Lysaght Australia started manufacturing corrugated iron in Australia and only a few years after that began exporting it back to England and to other markets. The material has evolved into Zincalume and Colorbond, more modern in style and more durable. BlueScope Steel still makes around 100,000 tonnes of the stuff every year and there's no end in sight, with the shed-building industry going rather nicely. In fact, they're not just sheds any more, but 'cold-formed steel buildings'.

To most people, of course, they'll keep on being sheds. They'll still sound good when it rains and annoying when a kid runs a stick along a galvo wall. Corrugated iron will be around for a long time to come. ▤

ABOVE Queensland shed window.
LEFT The corrugated iron water tank has 101 lives.

GARAGE MAHAL

A car collector's shed gets bigger and better, with more space for more cars and more projects until it becomes a sort of palace dedicated to our four-wheeled friends. It becomes a Garage Mahal.

George is the proud possessor of one such noble structure. It is impossible to do it justice in one photograph. Down the centre are his twenty or so cars, all in working order. Off to one side is a complete railway station for a 7¼ inch gauge model railway that goes for 750 metres around his historic garden. Down one end of the shed is a reconstructed country garage that would make the average oil-tin and garage-sign collector die with pleasure.

'It's a working museum,' says George. 'It's definitely not a sterile display. There's lots to do here.'

GARAGE MAHAL

One of the outstanding cars in his collection is the beautiful 1927 Bugatti 35C shown on page 143. George still races it on rare occasions. It can reach 130 km/hour — pretty good for a 75-year-old racing car and a driver almost as old. One of the complications of driving the Bugatti is that it requires a special pair of shoes with custom-made narrow soles. Ettore Bugatti, an artist turned carmaker if ever there was one, designed the space for accelerator, brake and clutch without sufficient room.

Another highlight of George's collection is the 1936 Mercedes Benz 540K Cabriolet B, a massive five-metre-long brute. This model, of which only about 400 were ever made, was one of the fastest touring cars of its day and had high prestige in Nazi Germany. This particular vehicle was used for some time by Professor

Ernst Heinkel, the designer of many of Germany's aircraft. It may also have belonged at one time to Luftwaffe chief Hermann Goering.

George acquired it at an auction after its owner attempted to bring it into Australia without declaring its real value. It had not clocked up many kilometres, but it had been neglected, and George had it carefully restored by specialists. The car has an air of huge and suppressed power about it: almost an embodiment of history in a vehicle.

George has always had a few cars but he started getting serious as retirement approached. 'Car collecting and having the workshop were a chance for me to phase gently into retirement. I love being here. I do a lot of entertainment and fundraising here for various charities.'

George has no trouble attracting visitors to his amazing garage. Kids get a huge kick out of the train ride, which crosses three trestle bridges as it travels through the historic garden established in 1873.

George has built his 'dream shed', and now gets great pleasure out of sharing it with other people. 'It's a delightfully incurable interest, long since past the hobby level, and I'm thoroughly enjoying the momentum it provides.'

THERE'S MONEY IN SHEDS

The 12 foot by 18 foot shed shown here is of great historical significance.

It's a backyard workshop at 369 Addison Avenue, Palo Alto, California. Here in 1939 Bill Hewlett and Dave Packard from nearby Stanford University started an electronics company that has since gone on to do rather well. In 1989 the shed was officially designated California State Historical Landmark No 976, the Birthplace of Silicon Valley. Similar landmarks dot this part of California. Nearby Los Altos has the garage where Steve Jobs and Steve Wozniak built the first Apple computers in the '70s. And internet search engine company Google — whose market worth is now US$125 billion — started in yet another Silicon Valley house and garage in 1998.

Keep trying . . .

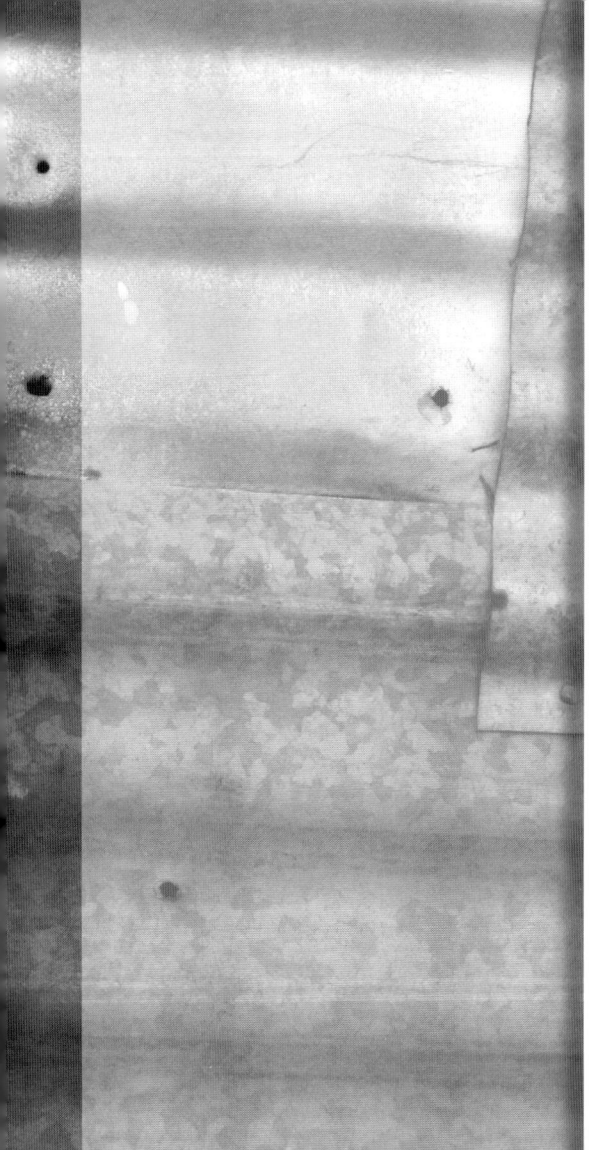

NOW THAT'S A BIKE

'What time is it? Hmmm . . . It's not too late to disturb the neighbours.'

Having given the neighbours due consideration, Greg plugs a heater coil into what looks to be a 6-inch-long jet engine on a temporary workbench. In fact, that's exactly what it is — a small gas turbine. Greg checks the switches, fuel leads and wiring, then starts a small battery-powered motor on the front of the turbine. A high-pitched whine fills the air. There's a brief flash of flame, and a growing heat haze appears out of the back of the turbine. The whine rises in volume and pitch and a familiar airport smell of burnt kerosene fills the shed. The whine reaches well over the 100-decibel mark for a few seconds — it's an awesomely loud high-frequency sound — before Greg eases the revs down from 30,000 rpm to idle for a few moments, then off.

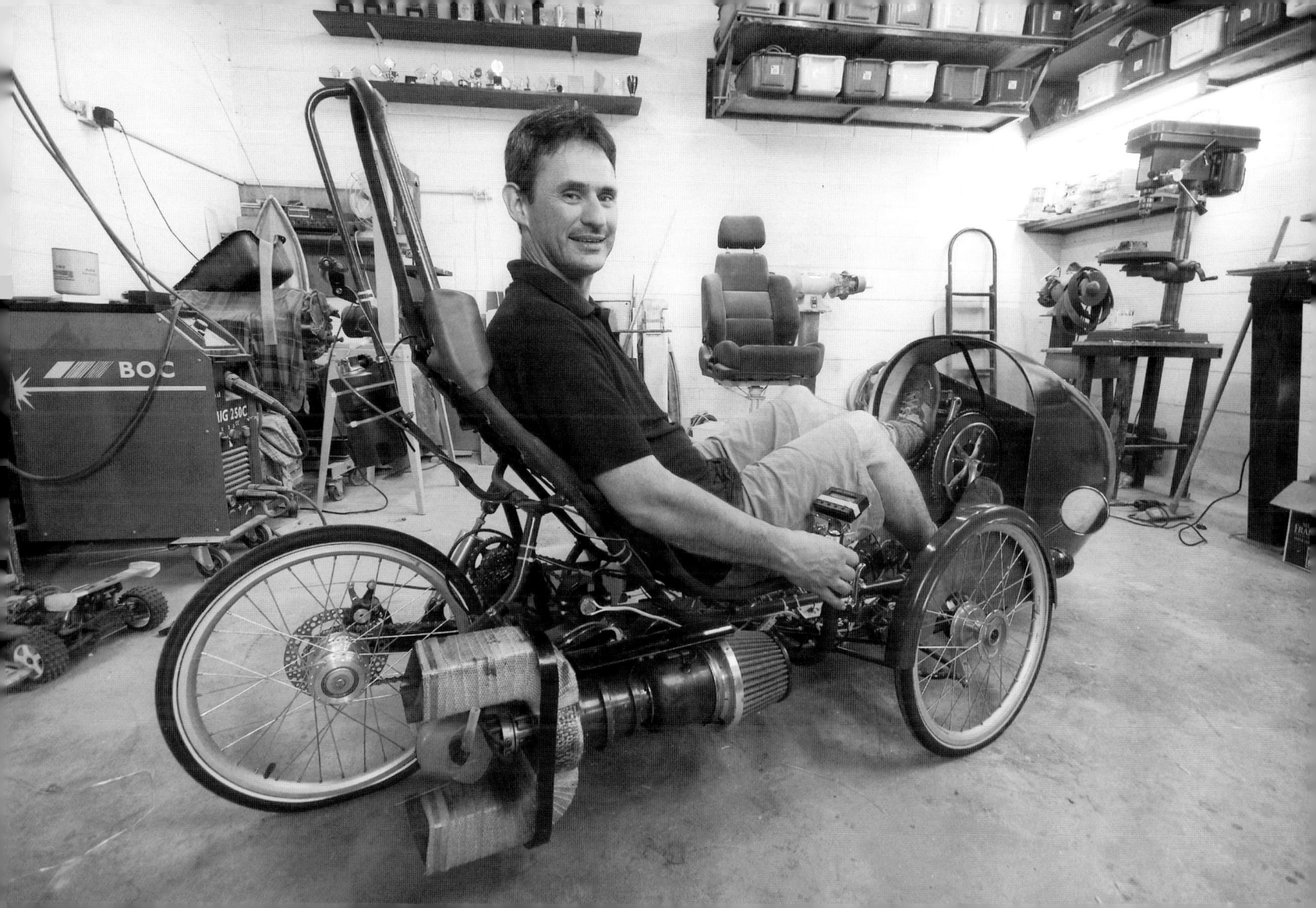

NOW THAT'S A BIKE

He does have to think of the neighbours sometimes but hell, it's impressive.

'It can go well over 100,000 rpm, which is quite a bit louder than that.'

Greg works as a technician developing specialised machinery for a company that makes industrial temperature-measuring equipment. In his spare time he's been building a gas-turbine powered recumbent bicycle, more or less completely from scratch.

It is possible to buy ready-made small gas turbines but Greg is interested in the challenge of making one.

The gas turbine works by forcing air and vaporised fuel (kerosene) into a combustion chamber and igniting the mixture. The gases from the contained explosion that results are directed over the turbine's blades, spinning the turbine and providing power.

It's quite a challenge to make a gas turbine in the shed as the combustion chamber heats up to about 650 degrees Celsius. In addition, because the parts move at such high speeds, any imbalance could make the whole thing vibrate and fail.

So Greg had to cast an accurate turbine wheel from Inconel, a nickel-based alloy that has a melting temperature of 1,200 degrees. In order to do this he had to teach himself investment casting, build a furnace and make a vacuum pump.

'There was a lot of trial and error, a lot . . .' he says, showing a box full of dud castings. 'It took a lot of patience. The first version of most parts didn't succeed but I just kept going. It's simple really, a question of fine tolerances and getting things right.'

While some of the information on how to build a small gas turbine is readily available on the internet, there is not a lot of direct or first-hand experience around to draw upon. So Greg spent a lot of time thinking about the design beforehand. 'I might do a few sketches but mostly I mull over it, running it over and over again in my mind. Then I go and make it.'

There are some admirable pieces of resourceful design in the turbine. The fuel rail is a circular brass tube with silver-soldered hypodermic needles coming off it so as to spray finely-atomised kerosene into the combustion chamber. The outside cowl is an old camping gas canister, and the reduction gearbox coming off the end of the engine to the bike's drive train was cannibalised from a cheap angle grinder.

Greg regrets not taking more photos

and movies of his sometimes spectacular early experiments. There were a few fireballs, he says, but mostly when things have gone wrong, there's been a fire inside the turbine rather than an actual explosion. He follows obvious safety rules.

The result of Greg's efforts is a bike that goes at least 60 kilometres an hour and generates eight kilowatts. It could go faster but he's happy enough going at that speed in a recumbent bike.

'It's cost me about $15,000, I suppose, but I don't have a problem with that. It's good knowledge to have and fun as well.'

It's a jaw-dropping, head-turning sight, Greg tooling around his quiet suburb at great speed on a jet-powered recumbent bike with the revs rising to a high screaming whine.

THE CREATIVE SHED

THE CREATIVE SHED

Creativity and sheds go hand in hand and it's important to draw a connection between the official creativity that is allocated to artists and the other forms of creativity that some of the other people in this book get up to in their sheds.

When you get down to tin tacks, artists are people lucky enough or smart enough to be able to muck around in their sheds all day long and make a living at it. Whether you are an artist in a studio or a car restorer in a shed, it's all making and tinkering in one form or another, with long passages of pleasant, dreamlike but focused activity interrupted by the occasional burst of frustration. Intermittently a conclusion is reached, a brief bubble of satisfaction pops and then you move on.

That's what it's all about.

Some artists have recently created works featuring sheds. This may be significant or it may mean nothing at all.

English artist Simon Starling won the heavyweight Turner Prize in 2005 for a work entitled 'Shedboatshed'. Starling dismantled a shed on the banks of the Rhine River and turned it into a boat, which he then sailed downriver to Basel, where he dismantled the boat and rebuilt it into a shed for an exhibition there.

A few years earlier Cornelia Parker, another English artist, produced a work entitled 'Cold Dark Matter: an exploded view' in which she enlisted the help of the British Army to explode a small garden shed. The fragments of the exploded shed were then suspended in the air.

She, too, won a prize.

Visual artists are not the only ones with an interest in the creative potential of the backyard-shed space. Writers have a long history of building and using small shed-like places to write the great novel. Here in Australia, the novelist Xavier Herbert wrote his great work *Capricornia* in a small

Queensland shed; numerous other more contemporary Australian poets and writers happily work in backyard sheds. In 1906, George Bernard Shaw, noted playwright and social commentator, built a writing shed which rotated on castors so that it could catch as much daylight as possible — an understandable innovation given the dreadful English climate. Mark Twain, Dylan Thomas and Ernest Hemingway were also known to have worked in backyard set-ups of various kinds. It is pleasing to report that Roald Dahl, the prolific writer both of children's stories (*Willy Wonka and the Chocolate Factory*) and adult fiction, worked in a grubby and chaotic garden shed. The British National Trust still has care of Virginia Woolf's writing shed at Monk's House, East Sussex, where she wrote *A Room of One's Own*,

which, among other things, spoke about the need for a personal creative space.

The need for such intimate creative space is not confined to the British or writers: Gustav Mahler, one of the last great classical composers, had a number of sheds in which he composed some of his greatest works. German philosopher Martin Heidegger also had *die Hütte* in the Black Forest in which he wrote *Building, Dwelling, Thinking*, a philosophical work about the nature of our relationship with built space: a suitable subject for a shed dweller. Sadly, he blotted the copybook by also being a Nazi.

There is some debate among the professionally creative as to how comfortable a work shed should be: some say it should not be too comfortable or have any distractions. Some even deny

themselves the small pleasure of a window looking out upon the world, which seems a little extreme.

Whatever the degree of comfort, chaos or character your shed may possess, perhaps the last word on the creative shed should go to Virginia Woolf, who spoke like a true shed dweller when she wrote: 'It is in our idleness, in our dreams, that the submerged truth sometimes comes to the top.'

TSOURAS, REBAB AND OUD

Musical instrument makers have some of the best workshop sheds you'll ever see. Perhaps it's the unusual tools and the exotic timbers, or the fact that these sheds tend to be small workspaces in which music sounds good. Or it could be that the task of making an instrument requires a certain passion that seems to permeate the space itself. Like many other musical instrument makers in Australia, Dimitri works alone in his backyard shed. What makes him different is the scope of his work.

The origins of the instruments Dmitri makes range from Greece east through Turkey and into the Middle East. Australians are generally familiar with the bouzouki from films such as *Zorba the Greek*, and as an accompaniment to transplanted Greek culture, but the bouzouki is only the tip of a very large musical iceberg.

TSOURAS, REBAB AND OUD

The oud, the tar, the rebab, the baglama, the tsouras, to name a few — Dimitri makes and repairs many instruments that are played in back rooms and small communities across Australia. They are part of rich and ancient musical cultures, which, encouraged by the rise of the World Music phenomenon, are starting to make themselves heard here.

Dimitri's beginnings were conventional enough. Having played guitar for some time he took up making guitars. He went to live in Poland for several years to work with violin maker Krzystof Mroz, and became much more skilled at his craft.

'Violin-making in Poland is still very traditional . . . but Krzystof was open and willing to show me everything he knew. I learnt a lot from him.'

With that knowledge under his belt he returned to Australia. At about this time he encountered rembetika, a form of music often described as the blues of Greece. The subject matter — life's sadness and the disasters of love, liquor and loss — are extraordinarily similar to the blues, as is rembetika's low-life reputation. All of this made the music very attractive to Dimitri and to quite a few other Greek Australians.

'When I first heard rembetika, I fell in love with it. It drew me in so strongly,' says Dimitri. 'So I started making some of the instruments on which the music was played such as the tsouras, a slightly smaller version of the bouzouki, or the baglama, a tiny version of the bouzouki which can be carried in a deep coat pocket.'

He studied photos of the instruments and researched as much as he could, eventually travelling to Greece to check out the makers there. Dimitri also started making the oud, the predecessor of the lute and guitar which has a very ancient history.

As he gained experience as a rare-instrument maker and repairer, friends and musicians such as Jeremiah (in the background), a widely experienced musician, started dropping in to talk shop and jam for a bit. Music that has its origins in the mountains of Afghanistan fills this shed in an Australian backyard as Dimitri and Jeremiah work out a complicated sequence.

'Making instruments is a lovely job — it's disciplined but creative work. And my shed is a good space for it but it can be very solitary. So I look forward to musicians and friends dropping in. There are times when people are playing that I think: there it is complete, the wood, the trees, nature, my work, all come together to create this sound we're hearing. We're all a part of it, with no beginning or end. These are moments worth waiting for, little reminders that we are heading in the right direction.'

FAR RIGHT A nearly completed oud. The oud is the predecessor of the lute and the guitar
ABOVE Instrument bowls of various degrees of completion. The end result will be delicate but surprisingly strong.
BELOW Final work on the soundboard of a bouzouki before it is assembled.

THE SATISFACTION OF STONE

'Michelangelo had twenty assistants to help him with his artwork, you know,' says Silvio as he prepares his compressed-air hammer. 'As for me, I've got one assistant and twenty horsepower.'

He is about to start the morning's work in the shade of the overhang from his large shed. All around him are blocks of granite, many carved into Australian wildlife themes: marsupials, sea creatures, birds and reptiles, invariably in accurate proportion and detail and having a beautiful glossy finish.

'You need peace of mind to work creatively and you can get that out here in the bush.'

Silvio hasn't always had peace and quiet. Sculpting in stone inevitably creates a certain amount of noise and dust, and his neighbours in the

THE SATISFACTION OF STONE

suburbs didn't always appreciate his efforts in the 8 x 15 shed where he started out. He now works in a huge iron building, complete with an overhead crane. He points out with some pride that he built his current shed from used iron scavenged from an old government shed — the only new things he used were the screws. 'Some of the screw and nail holes in the iron show that this is the fifth time it's been used. Built the whole place for tuppence halfpenny . . . but I suppose I've always been a scrounger.

'I started my sculpture career at five years old, when I pinched a block of soap and a kitchen knife from my mum to do a bit of carving. Later I graduated to plasticine.'

He started high school in the top stream but drifted downwards — which was an advantage from Silvio's point of view as he got to do woodwork. Although he was in the bottom stream by the time he finished high school, he got into art school, where he blossomed. He tried teaching and studied overseas for a while, then spent nearly fifteen years working for monumental masons. He started as an 'improver' (a tradesman who has not served a formal apprenticeship), and later became a contractor.

'I worked for twelve years at one place doing inscriptions in stone. I had to learn lots of industrial processes and the pay was based on results. I learnt to work very efficiently.'

By 1988 Silvio was able to work full time as a sculptor, and he has done so ever since. In his workshop are dozens of artworks being created using a wide range of methods and techniques: ground, moulded, cast, trimmed and polished. They range from the tiny frogs one centimetre across to a giant stone suitcase two and a half metres long. The impression is of someone very busy and organised. The two main rooms of the shed have a pleasing intimacy about them because the walls are covered in posters, drawings, other people's artworks and ideas: it's a bit like being inside a large busy brain preoccupied with many things.

Like many other artists' workshops, Silvio's shed is part industrial space, part very big playroom. Half his luck!

THE DESERT AND THE BACKYARD

Technically it's a shed but it's also a lot more than that. Sitting in the corner of their backyard, Hossein and Angela's studio is a large right-angled workshop where each can work separately.

'He's got his end and I've got mine,' says Angela. 'But there's this room at the corner where we meet. It's a good solution.'

Angela has been working in ceramics for many years and has a considerable reputation for her elegantly understated bowls and containers. Her pieces are in the muted colours of the desert. They are often unglazed or even sandblasted, weather-beaten and pitted to become earthen shapes of ageless simplicity. Lately she has been making more organic shapes suggesting casts of lost or never-known life forms. Her work and Hossein's have a similar preoccupation with time, land

THE DESERT AND THE BACKYARD

and space and they often work together. The physical arrangement of their studio helps.

'Creativity is a space where you can dream and play,' says Hossein. 'We've been very fortunate to have a place like this . . .'

It's a very different space to the one where Hossein grew up in Tehran. 'My memory of my childhood is that we had a tiny backyard, more of a courtyard really. When I was about 15 years old and decided I wanted to be an artist, the only space I could get at home was a small cellar, a low-roofed storage room. I thought I was very bohemian and smoked cigarettes in there.'

Not that his experience of space was all small and cramped. His family had lived in Baluchistan for some time, so he had known the desert. Hossein's artwork, which has a frugal sparseness to it, seems to echo the sense of the void that is common to his country of birth and Australia, his chosen home. Deserts have a long association with meditative states. Perhaps it's the stars at night in their uncountable billions or the baking heat of the long day: human beings are put in their place as tiny but precious specks in the scheme of things.

Hossein's part of the shed has an uncluttered, timeless quality to it that seems to speak of the desert. 'My work starts from a very small space, a state of mind and imagination that is very private. I aim to have the viewer dream themselves into that space. It's inevitable that it connects to other things and is not limited by my own experience.'

CREATIVITY IS A SPACE WHERE YOU CAN DREAM AND PLAY. WE'VE BEEN VERY FORTUNATE TO HAVE A PLACE LIKE THIS . . .

RIGHT Angela's work draws heavily on organic forms — it is difficult to tell the difference between the hand-made and the natural things around the work space.
FAR RIGHT Hossein also uses many organic objects — often twigs or small branches formed into letters or subtly turned into pleasing shapes.

MORE THAN MAKING DO

Even though he's one of Australia's foremost furniture designers, Khai still needs a shed. Nowadays it is a highly-organised workspace mainly used to work on prototypes and polish furniture but much of his early design and building experience happened in far more ramshackle sheds.

Khai's appreciation of Australian furniture began in the 1970s when he was working as a 'scrubber', stripping and polishing old furniture. When ancient chairs or wardrobes fell apart in the evil vats of hot caustic soda used for stripping at the time, the makers' construction methods were revealed. It was an excellent way to gain an understanding of how furniture was made — a sort of cabinet-making archaeology. As part of this busy little industry, there was just a glimmer of interest in primitive Australian bush furniture. Something about these old sideboards, cupboards, tables and chairs caught Khai's eye. 'These were simple,

MORE THAN MAKING DO

basic pieces of furniture, knocked up with available materials and often painted in oil-based enamels or milk-based paint,' says Khai. 'I liked their simplicity and integrity. They used local timbers, which gave them a certain quality, and whatever else was at hand such as corrugated iron, hessian, muslin, old linoleum and bits of leftover stringybark — anything. It gave that furniture a very distinctive Australian look. I've looked at frontier furniture from all over the world and each country has its own style, even though they all have that frugal simplicity in common.'

For Khai, going out into the bush to buy furniture from old farmers was a great experience. 'Travelling down the back roads was a good way to learn about Australia, to understand its social fabric. It gave me a respect for how people lived and "made do" out there.'

According to Khai, the broader appreciation of this kind of furniture didn't take place until the 1988 Bicentennial.

'At that time there was an awakened appreciation of things Australian. You also had a number of big collectors such as Lord Alistair McAlpine rapidly buying up everything that was going. So you had furniture that the Salvos wouldn't look at suddenly being worth huge amounts of money.'

It's now quite hard to find original bush furniture. Original Jimmy Possum chairs from Tasmania, for instance, fetch large sums and are displayed in state art galleries. Khai, for his part, has evolved a design style of his own based on frugal bush furniture, on the furniture of his Asian background, and on classical European design. 'I'm picking and choosing from everywhere and blending it, chucking it in the melting pot. We still have to "make do" in some ways because there is no great established tradition here, no set guidelines or established infrastructure — whether it is in how we design things, make things or market them. You have no choice other than to be independent.'

Khai's distinctive and exquisitely-made chairs are an elegant reminder of the strength of that independent streak.

RIGHT Khai's take on a bush classic — the Jimmy Possum chair.
FAR RIGHT French polishing is one of those activities that use arcane materials and strange liquids.

8
9

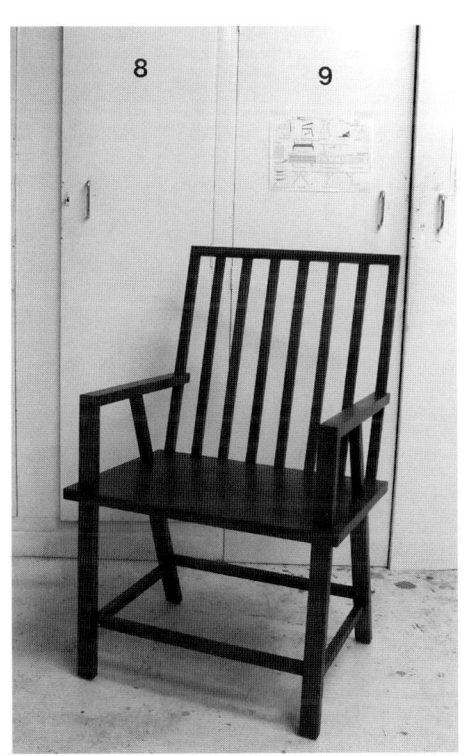

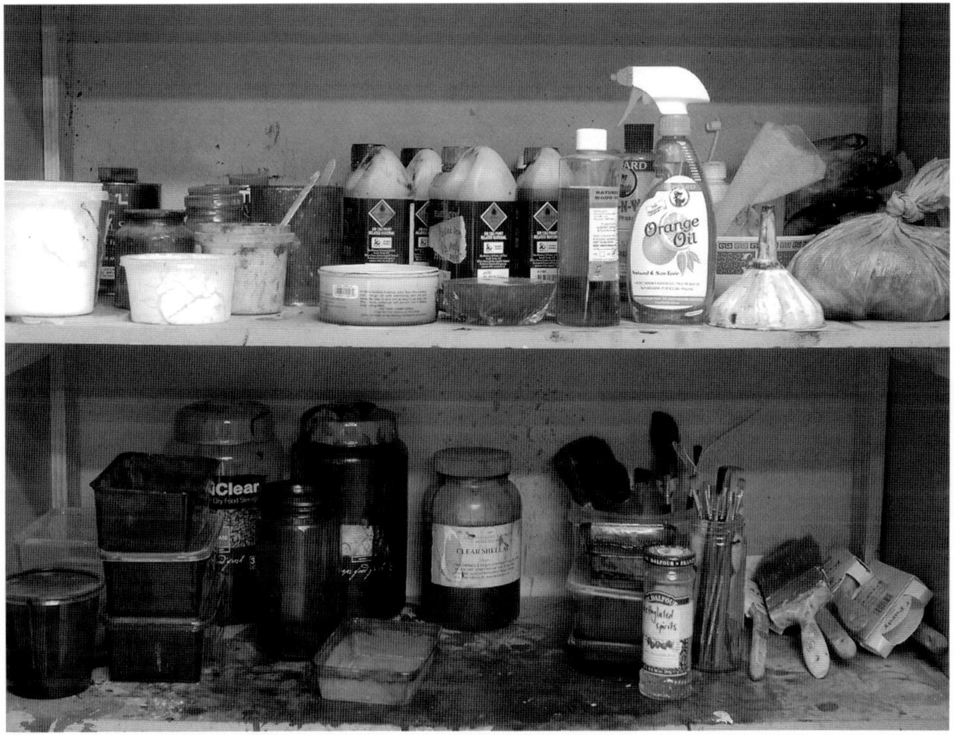
Orange Oil
CLEAR SHELLAC

RESONATOR TERRITORY

A few years back Don took the momentous step of leaving his safe but mind-numbing job in the government to make guitars for a living. He had got stuck in a dull clerical job as part of a classic rock 'n' roll story: spend your raucous youth playing to drunks, driving around in old vans, lugging gear in and out of pubs in the middle of the night and generally leading the high life; then suddenly you're in your 30s with not a lot to show for it, and your friends who became schoolteachers are putting money on a second house; so it's into the government for the ageing superstar. Apparently the Australian Tax Office could have put some excellent blues and rock bands together from their employees with that kind of history.

But the long, long haul for the superannuation payout was not for Don. He experimented with making steel-bodied resonator guitars.

THEY'RE THOSE SHINY METAL GUITARS OFTEN ASSOCIATED WITH BLUES SLIDE GUITAR — THEY HAVE A METALLIC AND NOT PARTICULARLY SUBTLE SOUND, BUT IN THE HANDS OF A FORCEFUL PLAYER SUCH AS DON, THEY RING LIKE THE BELLS OF HELL.

They're those shiny metal guitars often associated with blues slide guitar: they have a metallic and not particularly subtle sound, but in the hands of a forceful player such as Don, they ring like the bells of hell.

In starting up his own brand of resonator guitar — 'Donmo' — Don found a niche and he was soon getting orders over the internet. He experimented with variations including combinations of wood and metal, and recently a successful (and impressively loud) resonator mandolin.

Then he made a breakthrough that should endear him to shed scientists around the nation: he put an old piece of Lysaght corrugated roofing iron through his flattening roller and made a resonator guitar with it. It looked dinged up and ancient but it sounded terrific. In fact people loved the instant crusty look of

this type of guitar and Don has plenty of orders.

So Don is hard at work making guitars from sheds in his father's old shed. The blues and sheds seem to just fit naturally: perhaps it's in the essential nature of stripping back life to its core. It seems just right to have an instrument made from such a characteristic Australian material.

Don is on to something big here.
www.donmo.com

RIGHT Lysaght corrugated iron gains a new and undreamed of life.
FAR RIGHT Donmo resonator guitar at the place of birth.

DEFYING GRAVITY

As he is one of the few professional traditional chair-makers in Australia, you would expect Howard to have a pretty good shed.

It doesn't disappoint. It's slightly ramshackle and worn around the edges and it's snug, even in the freezing cold winters on the edge of Barrington Tops in New South Wales. The galvo roof and walls are punctured with nail holes from a former life, so that any photo taken inside the shed is full of tiny white dots.

Howard is no slouch in the photography business himself, having been a professional for many years. Some time ago his interest in chair-making, in particular the traditional Windsor-style chair, convinced him of the need for a career change. Since then he has been practising, teaching and demonstrating chair-making from scratch, without the aid

of power tools and using mainly Australian timbers.

Windsor chairs are quite a cult in their own right. To the untrained eye they might look like an Olde Worlde notion of a chair, solid and respectable. When you look at how they've evolved from their 17th century English and then later American influences, you can see how they have become slimmer and more flowing. They are delicate yet strong, and extremely comfortable. Sit on one of these chairs and you'll feel better immediately.

At the end of ten days, everyone who participates in one of Howard's classes goes away with a chair which, depending on the individual's feel for working with wood and tools, can suffice to keep a human body off the ground or be a dazzling example of the chairmaker's art. Along the way they will

have used a range of tools and methods rarely seen anywhere else, such as the pole lathe, a treadle-operated device that uses the natural springiness of a sapling to turn the wood.

The fact that Howard can coax a result — a chair you can sit on — out of people with a very broad range of aptitudes is a great testament to him as a teacher. He can light a fire in people, even those convinced that they're 'not good with their hands'.

Making a chair is a personal moral act. A good chair dignifies human life. It keeps the human backside off the ground and it's a piece of engineering in its own right. Joining four narrow sticks so that they can take the weight of 100 kilograms or so of human and do so for, say, 100 years without wobbling, is no small achievement.

Howard's shed is purposeful and

functional, yet it has the air of a place where there has been joy and a moment or three when blinding insight into the ways of wood has struck like lightning.

'In these days of rampant technology, to work with traditional hand tools gives pleasure to the soul,' muses Howard. 'To work at your own pace and to stand back at the end of the day and admire your achievements must give you some insight into the very roots of woodworking and thus to the ancient relationship of humankind and that miracle we call wood.' www.rarechairs.com.au

THE BRITS DISCOVER SHEDS

Since the original publication of *Blokes & Sheds* in Australia, a number of similar publications have sprung up elsewhere in the world. In Britain, *Men and Sheds* highlighted the backyard or allotment shed and brought about a frenzy of activity, especially on the internet. Allotments are not common in Australia but have existed for a long time in the UK. Public land is set aside for use as vegetable gardens for people without the 'luxury' of substantial backyards that we have had, until recently, in this country. Many allotment tenants will erect small sheds out of scrap materials. The British backyard shed is generally a wooden structure of very modest proportions: to the Australian eye, it bears a strong resemblance to a slightly enlarged dunny.

The UK shed has many websites devoted to it (such as www.readersheds. co.uk) and the sheddies are working on a series of high profile promotion activities such as 'Shed of the Year'. They also tried to have a 'National Shed Week' declared by the government (not much luck there, curiously enough) However, a recent survey did find that the backyard shed was considered the favourite British institution by a sizable number of respondents — it was even more popular than cricket, which may not be surprising. What was more surprising was that some 30 per cent of the respondents under 25 had had sex in the family shed (away from the parents). Only four per cent used the shed as a workshop!

The sort of attention that sheds have been getting in the UK has led to the inevitable gentrification of sheds or, as they are now becoming known, the 'garden office'. The designer shed is all in the best possible taste and is catering to the ever growing number of people working from home, usually working in internet-based businesses. Some promoters of these 'Garden Offices' claim that their constructions add as much as £25,000 to the value of a home.

Despite the fact that the shed has limited appeal in the UK for climatic reasons — you could freeze to death in an unheated, uninsulated space for a fair part of the year — there are still plenty of people who have grasped the fundamental virtue of the shed. One of these is Scotsman Bob Davidson, who has written a fine song entitled 'I'm gonna live in my shed' which you can hear on his website (www.bobdavidson. co.uk). It's worth a listen.

ON THE TOOLS

Chapter 6
ON THE TOOLS

A good shed needs tools: the people in this chapter have a few things to say on the subject. Earlier we discussed the bush-mechanic tradition of temporary repair which can become permanent repair. This is regarded in some quarters as a terrible disease that infects the working population with poor standards of work. That view is not borne out by the facts: the people in this chapter have standards that are far from rough and ready. There is in Australia a small but healthy sector of tool collectors, tool connoisseurs and tool makers who uphold very high standards indeed.

The sorts of tools they collect are, by and large, not machine-driven tools but hand powered. While we may appreciate the convenience of the battery-powered drill or the brute force of the angle grinder, these are not tools which one can imagine tool collectors drooling over in a couple of hundred years time. Hand tools have a history to them: an Egyptian carpenter of 4000 years ago would have no trouble recognising and using, for instance, a handsaw, although he might be surprised at, and certainly appreciate, the sharpness and durability of modern steel in the blade. Old tools also acquire a seductive quality.

Hundreds of hours of use somehow seem to imbed a human-scaled quality to the tool: it may be in the patina acquired from linseed oil rubbed into the handle and heated by hard-working human hands or the dull polish of old steel well looked after. Even better, it is possible to buy very old tools which simply need sharpening and still work perfectly well and sometimes even better than their cheaper modern equivalents. It's very satisfying to feel a newly sharpened chisel slice its way cleanly and efficiently through timber. No screaming power saw can ever hope to match the subtlety of that feeling. The Italians, with their ancient artisan traditions, have an excellent expression for power tools: *il cretino veloce* which translates as 'the fast idiot'.

Although tool collecting has accelerated in recent years it is still possible to pick up some very good quality tools at garage sales, junk shops and clearing sales for very little money. This may not last long though as the depth and value of the mundane instruments of skilled manual trades are once more being appreciated. As skill shortages grow, those skilled trades are looking more and more like the ruins of a great culture revealed by a lowering tide. There is every chance that we will see a renewed respect for what was once termed dismissively 'working with the hands', even if it is a respect based on the fact that we'll have to pay a premium for that work. That's the price of our collective lack of foresight.

But it is wrong to talk about the people in this chapter as 'working with the hands': no-one here works only with their hands. They work with their hands, and with the rest of their bodies. Until we acknowledge their wits and creativity, we will not appreciate what they have to offer nor address the question of skill shortages.

That's what tools are all about — more power to the hand.

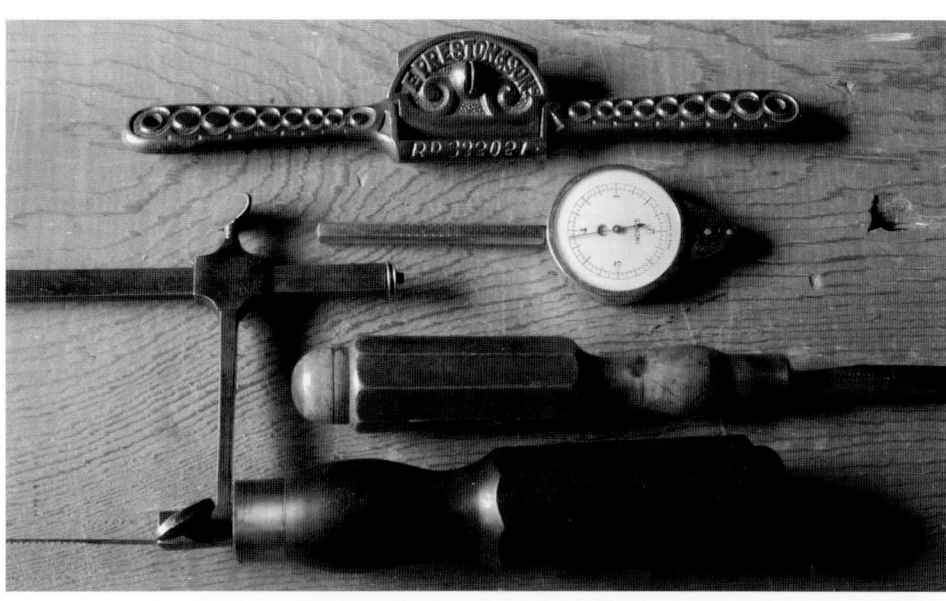

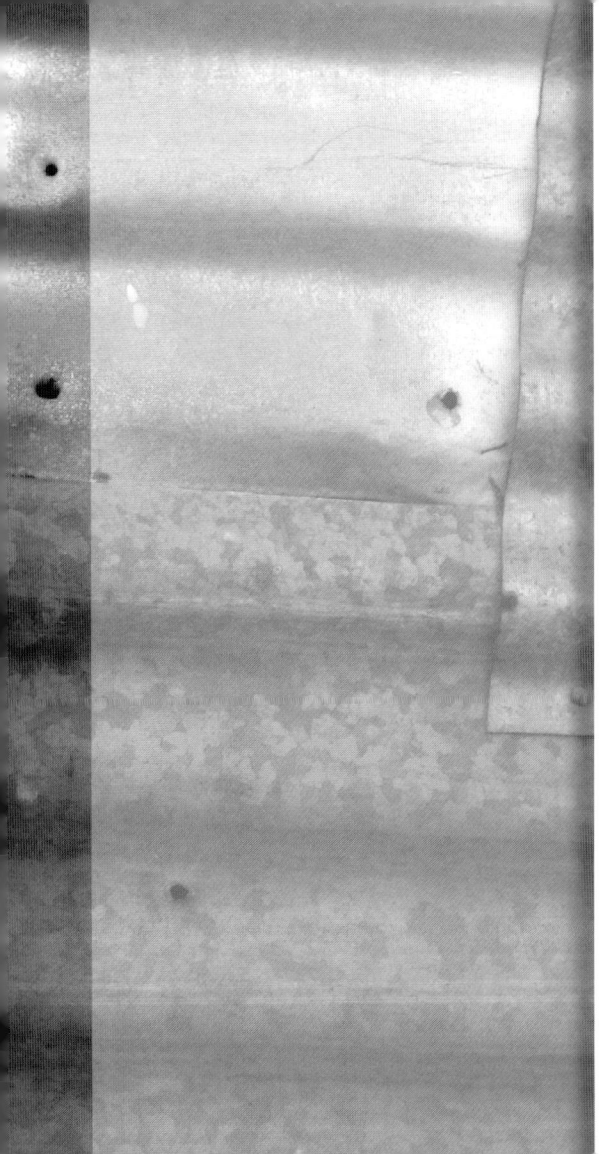

TOOL-AID

Any shed worth its salt has a good set of tools. Kees has a few more than most in what is closer to a cave of treasures than a shed. Nearly every surface is covered with hand tools, all of them excellent examples of their kind and most in impeccable condition.

It's not hard to understand why people become keen collectors of old woodworking tools. These tools age in individual ways, acquiring a pleasing shape and patina from the hands of their owners, sometimes multiple generations of them, yet are often still effective working tools. They still have the power they were invested with at the time of their manufacture, even if they date from the early nineteenth century. Old tools such as these are part of the history of modern technology.

Kees is one of those passionate collectors. He spends a lot of time

maintaining the tools he's collected over the past 35 years. His shed, his collection, is a pleasant retreat centre for him.

But Kees's interest and expertise in tools has taken him much further afield than his shed.

A few years back when East Timor gained its independence, Kees and fellow members of the Hand Tool Preservation Association of Australia joined with Freemasons Victoria and The Knights of the Southern Cross in offering a hand. They knew that many East Timorese had few resources such as tools, and scarce access to electricity. At the same time, Australians were buying cheaply made power tools such as circular saws and power drills in ever greater numbers — and neglecting or discarding perfectly

functional hand tools such as tenon saws or brace-and-bit drills.

It seemed like an opportunity to do some good.

They put out a call in the media for old tools — and were overwhelmed by the response, receiving some 3.6 million items. Among the tools collected were some rare and valuable pieces. These were sold and the proceeds — about $18,800 — used to buy new power tools for East Timor. Some tools were not up to scratch or were not suitable for use, but a large number were cleaned up, sharpened if necessary and packed into shipping containers. It was a huge job for the volunteers.

When the tools arrived in East Timor, it turned out no-one knew how to sharpen the saws. So a decision was taken to send Kees, a former technical teacher, over

to teach people how to sharpen saws and maintain tools. It was an interesting experience for Kees and his wife Nelly, seeing how people lived with few resources, little food and often no electricity. Many of the tools were being used in technical schools or in small villages. Although the project didn't cover as much of the country as hoped, it still did have an impact.

'You can't just barge in with this sort of project. You have to feel your way in. Some surprising things too . . . such as wood being scarce. But generally the tools were very well received and appreciated and put to use very quickly. If we go back, we'll be more prepared . . .'

TIPS FOR THE WOULD-BE TOOL COLLECTOR

Dave Whyte has been a professional tool dealer for many years and has an immense fund of knowledge on many curious and wonderful tools.

'It's hard to say just what motivates tool collectors. I think it's a fascination with technology. If you come face to face with a Norris A1 jointer plane with dovetail joints and rosewood infill, it's a beautiful tool made with a degree of excellence and it still works superbly, even though it was made somewhere from 1890 to 1910. It's not just woodworking tools that people are drawn to but also medical, scientific, blacksmiths' and mechanics' tools.

'If you're interested in collecting tools, garage sales and Sunday markets are probably the best place to start. It can be a bit hit and miss. It's a good idea to join something like the Hand Tool Preservation Association of Australia (HTPAA) and also get some reference material about old tools. From there you can go to auctions, which often have some refined collections, specialist tool dealers like me or specialty antique shops. The internet is also very useful. But the best way to start is just to go out and start looking.'
www.gizmotools.com

USE IS BEAUTY

For such a young bloke, Chris has managed to build up a very impressive shed/workshop. He's one of the small number of woodworking toolmakers in Australia and he started young.

He was bitten by the woodworking and fine-furniture-making bug at about the age of 15. Self taught, he read a lot, sought advice from experienced craftsmen and took on some big projects. Problems with the quality of the tools he was using led Chris to make some of his own, which in turn generated interest from other woodworkers. He has been making and selling tools since he was 18 years old.

Chris left school at 18 and did two years of a fitting and turning apprenticeship. He makes no bones about not being formally qualified, having preferred to work in the area between metalwork and woodwork,

USE IS BEAUTY

using experience and knowledge gained from both. The use of metalworking tools allows for precision, accuracy and sharpness — a sought-after quality in the beautiful gauges, bevels, squares and knives he makes. 'Here,' he says holding up a part of a machine he designed and made, 'that's my résumé.'

Though Chris maintains high standards of precision in his work, some of the machinery he uses is quite old. He's a keen historian of vintage industrial machinery and frequents sales and auctions of old equipment. He takes a purchase back to his shed, strips it down, preserves and restores it. He draws on the knowledge of older engineers who used and maintained this machinery when metal engineering was in its heyday. It is not widely appreciated just how large a machine-tool industry there

was in Australia just after World War Two. That industry has all but disappeared as a result of cheaper imports, which do the job adequately but leave many older tradesmen discomforted. Those old men take heart when someone like Chris takes a deep interest in their trades.

Chris's pride and joy is what he calls the bandosawrus — an ancient bandsaw, probably dating back to about 1870 and weighing about three-quarters of a tonne. He has restored it to immaculate working order. 'I use it nearly every day and it has no trouble sawing through timber up to 13 inches deep. It's a pleasure to use.'

Woodworkers are generally fascinated by timber and its qualities, and Chris is no exception. Timber for the handles and other parts of his tools — much of it Tasmanian blackwood, chosen for its fine

RIGHT Chris's cutting and marking tools have a balance and 'handiness' to them.

figuring — is stored in correctly humidified conditions in a dedicated part of his shed. Chris has a range of other timbers set aside, much of it unusual and obscure such as some 5,000-year-old black red gum, a dark-as-night hard timber that has spent most of the past five millennia buried underground.

Chris's tools all have an excellent sense of balance to them, the result of an endless process of evolution and constant improvement. 'I see it a bit like that old saying: if you're trying to make something beautiful, it may end up ugly — but if you're trying to make something useful, it can end up beautiful.'
www.chris-v.com

RIGHT Like all serious woodworkers Chris keeps a good store of quality timber in a climate controlled room.

MY HAPPY HEART SINGS

Attilio is a 'tornetorre', a fitter and turner, who has spent a lifetime combining the great Italian tradition of metalworking with an equally magnificent tradition — music.

His main metalworking shed (more about the others later) is where he still turns out shafts, pistons and similar items. His drums sit in the corner, ready for the next cabaret gig.

Old gears are piled up against walls featuring photos of cabaret acts from the 1950s and other musical paraphernalia. At the drop of a hat Attilio will burst into a bit of Dean Martin or perhaps a bit of light opera, delivered in fine bravura style.

'When I came out to Australia in 1954, I started a band and began shiftwork in the same week.' Attilio worked as a fitter and turner, first

MY HAPPY HEART SINGS

at a lawnmower factory and then at Chrysler's. He enjoyed the challenge of the work, which often involved complex manufacturing machinery — making replacement parts and generally keeping the show on the road. What was work is now a hobby. He still enjoys the precision of it.

HERE IS SOMEONE WHO KNOWS HOW TO LIVE AND DOES SO AT FULL THROTTLE, BECAUSE . . . WHEN YOU STOP, YOU'RE IN THE GROUND.

A similar resourcefulness went into making music. He spent five years singing in church but he was keen on learning to be a drummer. Drum kits were scarce in post-war northern Italy so Attilio made his own. He got the skin from a dead donkey, treated it with lime, stretched it over a large sieve and — hey presto! — he had a bass drum. The rest of the kit was created in similar style.

He and his band travelled to other villages, with their equipment festooning a bicycle.

They played the hit tunes of the day as well as tangos, cha-chas and later the modern and exotic samba. He played the same material once he came to Australia. For around half a century he has been comparing the moon to a big pizza pie — or sometimes a river. He still sings a couple of times a week, usually with Italian friends whose musical history is as long as his.

He has another shed. At least he calls it a shed — it's more of an open-sided pavilion with a big table and an oven and barbecue down one end. Here he lunches daily with his brother Joe and a friend or two, often enjoying a glass of their excellent home-made wine (the cellar is in yet another shed). There's a bit of singing and talking about the good old days. There's a bit of pasta, some vino and maybe a little espresso coffee 'corrected' with grappa to finish.

With one shed activity or another, it's a busy life. Here is someone who knows how to live and does so at full throttle, because as Attilio says, when you stop, you're in the ground.

ABOVE Although he is retired Attilio still does the occasional turning work.

RIGHT Hey mambo! Any chance to perform, Attilio will take it.

THE TRITON WORKBENCH

As a young ABC journalist in the 1970s, George bought a house in the 'burbs and fitted it out with the latest in beanbags and what few mod cons of the day he could afford. He got friendly with his neighbour Len, an antique-furniture restorer with a shed full of woodworking tools. Len saw that George had an aptitude for working with wood and suggested he build a good dining table as a project. 'Len said, "God knows you need the furniture."'

So George lashed out and bought several hundred dollars worth of cabinet-making grade timber — a lot of money at the time. But before he got around to starting work on the table, Len moved out, leaving George with an empty shed next door and a pile of expensive timber.

Not having any tools worthy of the name, George purchased a 9¼

THE TRITON WORKBENCH

inch Skil circular saw, which was just affordable. His first cuts in the expensive timber were shocking.

'I thought there must be a better way. If Len was here, I'd use his radial arm saw. I've got the motor and the blade — all I need is to make up a simple guide. A few bits of angle iron on an old table fixed that up. A bit later I needed a table saw, then a rip saw. I thought, "Hang on, if I set the saw up right, I could also cut rebates and tongue and grooves."'

Thus was born the idea of the Triton Workbench. For George it was the first step down a long road filled with many tribulations and, ultimately, triumph.

'I didn't set out to invent anything. I just wanted to salvage my money and pride. I couldn't believe I had wasted all that money.

'So I had this idea of turning a circular saw into a docking and rip saw, to make it easily convertible from one mode to another. The challenge was to make it adaptable to all saws, to make it affordable and to have good space-saving aspects. If I could get it right, it would be a handyman's dream.'

George made a number of prototypes, ran them past friends and tried to get people interested. He approached some hardware stores but they were not interested. 'The big tool-making companies said, "Gee, George, nice idea, but . . ."'

Finally, George took a punt and decided to make a hundred Tritons and sell them himself — anywhere. He asked a struggling engineering works to make components. 'I said to Don and Vic, the owners: "Teach me what I need to know. I'll do any free labour

you can use and I'll pay $4,000 upfront. Is that a fair deal?" They said, "No, it isn't, but we'll take it anyway." They're still Triton's number one supplier!'

George's house filled up with parts ready for assembly. 'I had to do all the grunt work — assembling, wiring, packaging. As bits arrived, none of them fitted together. My engineering drawings were shit.' George trudged on, working at his journalism job during the day and assembling workbenches at night.

In July 1976, as a last ditch effort, he went on the ABC's *The Inventors*. The next morning his phone went crazy — no sooner would he put it down than it would ring again instantly. By 3 pm he had 1,000 benches on order and within a few days three sacks of mail had turned up with more orders and cheques.

'I had orders, a public profile and cash, so I leapt in at the deep end of a very deep pool. It was an excruciating time of my life. Everything I touched turned to shit. Everything.

'Commercialising inventions is a minefield and I think I stepped on every mine, but I was able to keep going because I had a full order book. I was able to practise on my first unfortunate customers. The idea was a good one but the execution was woeful. What I should have done was hire a professional to refine the design rather than keeping the clunky user-unfriendly attributes of the first machine.'

Several versions later the Triton had improved but there were still problems.

'Everyone who wrote to me with problems or suggestions got a letter signed by me explaining that the idea was still being developed. That acknowledgment by people that I was doing my best saved me, I think.'

By 1981 George finally had something he was proud of — the Triton Mark 3. The business doubled every year throughout the 1980s.

Triton has since become a massive business which markets in 17 countries. George says that if there is a secret, it's that he has honoured the customer. 'We've treated them the way you want to be treated. For example, we've made new products retro-fittable so that old customers aren't forced into buying new stuff. We want people to get real satisfaction out of using our tools.'

George is no longer the owner of Triton Products but he is still the force behind the brilliant idea of the Triton Foundation, which gives objective advice to inventors. (If you're an inventor or would-be inventor, check the website www.tritonfoundation.org.au)

His current life in the rainforest allows George to explore his aptitude for working with wood. He has all the Triton products he could ever want, and takes lessons from master cabinetmaker Jeff Hanna of Lismore.

Asked if there is any way to sum up the success behind his innovative life, George recalls the favourite saying of '60s television scientist Professor Julius Sumner Miller: 'Why is it so?'

'The questioning aspect is my ace in the hole. There must always be a better way. I refuse to accept that things couldn't be done better.'

THE BENCH

'Tabula ipsa loquitur' is written in large letters high up in Mike's shed. It's Latin for 'The table speaks for itself'.

Mike's benches do speak for themselves. You can tell at a glance that they are balanced, functional products, pleasing to the eye and ready for work. In fact, Mike suspects that some people buy them for mere decoration: if you set up the bench in the lounge room, he observes, it should be easier for your partner to clean up the shavings at the end of the day.

Some people use workbenches made from old doors standing on trestles, sometimes for years on end. That's fine, but a good workbench can make life in the shed a lot easier.

Mike gives the following advice: 'A good workbench is flat and stable,

and will stay put. It's got to be heavy and not wobble. When you're sawing on it, a bench shouldn't move around — if you're chasing your bench around the shed you're wasting energy.

'The rule of thumb some old bloke told me is that a bench-top should be about ball height. This was true for the old joiners who spent long days hand-planing large pieces of wood and using moulding planes. These days we have our benches a little higher. Remember that the bench height is your working height and you should be able to stand up straight for most of your work. If you have to bend slightly to reach your work it's hard on your back and will shorten the time you can spend at the bench. A bench is a bit like an engineer's plate — everything you do has to come off that reference.'

There are other considerations to give to your bench, such as whether it is situated in good natural light so that you can see what you are doing. A gloomy dark shed is all very well if you are breeding spiders but if you are expecting to cut wood with any degree of accuracy then light matters. Another consideration is how you can get around the bench — ideally it is freestanding on all sides and has enough space to deal with some good lengths of timber to build a small ship, or whatever takes your fancy. Mike's benches form a substantial foundation for such activities. Mike's benches, which are usually made of silver ash (*Flindersia schottiana*) or mountain ash (*Eucalyptus regnans*), are hand planed from rough timber to a glassy smoothness.

Mike is one of those woodworkers whose wood planes are so sharp that the timber comes off in one long, even, transparent shaving. A quick check with the micrometer shows that the finishing shavings are only half to 1 thou thick. Impressive.

'I always get people asking why I don't make the benches cheaper. My answer is: quality of product and quality of life. Time spent making benches is my woodwork time, and I would much rather make quality pieces than bolt together a few old pallets.'

Living and working near Byron Bay does cause you to give quality of life a thought. You could live in a shed, just about any shed, here and life would be OK. From the door of his shed, Mike looks out over valleys and mountains covered in deep, dark rainforest. Half your luck, mate. www.michealconnorwoodwork.com.au

A GOOD WORKBENCH IS FLAT AND STABLE, AND WILL STAY PUT. IT'S GOT TO BE HEAVY AND NOT WOBBLE. WHEN YOU'RE SAWING ON IT A BENCH SHOULDN'T MOVE AROUND.

THE STUFF QUESTION

Sheds have traditionally been associated with the question of stuff, in particular useful stuff. Even better is useful stuff for which you have paid little or nothing.

The big question is how all this stuff is to be stored.

Broadly speaking there are two approaches to storage. First is the Random Access Method (RAM), also known as the dynamic random chaos approach, in which things are left as they were put down on arrival or at the end of the last job. The visual effect is as if a tiny cyclone has passed through the shed, but more of that later.

By contrast, there is the Disciplined Order System (DOS), under which objects are sorted and classified according to any one of a range of possible qualities: size, function, colour, material, application, and so on.

The latter approach allows for the making or obtaining of large amounts of storage stuff such as shelves, drawers and jars. Hours can grow into days and months in sorting stuff into these systems. A box from a garage sale is tipped out onto a workbench and the contents dispatched to nooks and crannies all over the shed. Those washers go into the section marked washers, subsection metal, subsection medium. That broken chisel is stored in the chisels to be sharpened box, maybe to rest there for years. This old piece of string — is it too good to throw out? It's a tough

RIGHT There's always someone with more stuff than you. Show these photos to someone that says you've got too much stuff. The location is secret.

call. Into the rubbish bin it goes. In a short time, all that remains of the box of garage sale goodies, apart from a big warm glow, is a pile of greasy sawdust and some bent nails. And there are people who happily straighten bent nails.

While the disciplined approach clearly has its own satisfactions, the dynamic random chaos approach has its benefits too.

The jury is still out as to which is the better method, although both can bring on the highly valued **stuff dream state**, where time is suspended as the shed scientist tinkers away happily. There are plenty of people who work effectively in what appears to be a concentrated pile of hurricane debris; they can remember where they left things, or there's a secret logic at work known only to the shed's occupant.

It keeps the brain ticking over nicely, you know. The obvious benefit is that you get things done, that the project is under way and happening. All too often under the disciplined approach, nothing is actually made. The work space may be absolutely perfectly ready for action but there's a slight disinclination to actually make a mess because you'll only have to clean it up later. However, the chaotic approach

can reach a certain point, known to shed scientists as terminal shambolification, where the normal rules of law and order have broken down. At this point work stops completely, usually to look for a drill bit or a chuck key. As with the disciplined sorting process, hours soon turn into days in an attempt to impose order . . . or at least head somewhere in that general direction.

GOOD SHED STUFF AND CRUMMY SHED STUFF

In the interests of advancing either system of classification, it's important to discriminate between good shed stuff and crummy shed stuff. In today's society, many of us feel that we're drowning in a sea of shoddy stuff, that things tend to be made with lighter gauge or inferior quality metal or from materials that lack substance.

Tools no longer have integrity: integrity as in the opposite to disintegration.

Good stuff has a permanency about it, a reassuring quality. It's not just a question of a tool or object being old: it should feel substantial enough to carry meaning. It may be a matter of rubbing back rusty or tarnished metal to reveal names like Disston, Millers Falls, Preston or Mathieson. We're talking about stuff

with a clear purpose to it. Even a wood screw has a mission in the world — to go into a piece of wood and hold on to it for dear life.

So sheds collect all this stuff because you know it will last and may yet see a whole new life as something useful, effective and valued — though maybe not in your own lifetime.

A LOT OF BALLS IN THE AIR: IN DEFENCE OF PROJECTITIS

Stuff comes in handy for projects. Some stuff may be a project itself, such as a machine, tool or piece of furniture waiting to be restored. Or the stuff could be something, like a left-handed widget stretcher, that's sure to come in handy for a project some day.

There has been some talk lately about a condition called projectitis.

Sufferers have multiple projects on the go simultaneously — some almost finished, some puttering along at a reasonable rate; their 'backburners' may be vast.

There is no known cure or treatment for projectitis, and there is a great deal of vicious discrimination against those afflicted. They are seen as unrestrained generators of disorder, and often cursed as jumblemucks, rumplemuddles or higgledy-pigs.

Of course, it's ludicrous to call projectitis an illness at all. It's more a form of gambling. Even if you finish only 10 per cent of your projects, you will have accomplished a lot more than someone who has never started out for fear of failure or of contracting projectitis.

Projectees (as we'll call them) are, in the end, supreme optimists. Even those with inflamed or chronic projectitis, who would need to live to 450 years of age to finish everything on their books, just keep ploughing steadily ahead.

Such optimism should never be crushed with mere concerns about so-called order. Aynsley (see Chapter 1) offers one of the best defences against the 'too much stuff' accusation: 'What's the problem? — this stuff's not eating anything.'

ABOVE More stuff. This is Richard's place (know as 'Magnetic' to the locals). If you need lots of stuff in a hurry, check out a farm clearing sale and buy everything left over at the end. Make sure you bring a big semitrailer.

THE SHED AS A PALACE

Chapter 7
THE SHED AS A PALACE

The shed is a highly personal domain. The success of the first *Blokes & Sheds* book inspired many shed owners and would-be owners to expand their activities. Sheds became legitimate . . . well, at least a bit more legitimate. Sheds are above all else places where men choose to define themselves by what they make and do — or perhaps plan to make and do, because what the shed could be is as important as what actually happens in it.

In a shed you have a chance to make something that is unmistakably your own, blunders and all. You might turn a fully-grown tree into a set of beautifully handcrafted toothpicks, but they will be your toothpicks to display casually to friends and family saying, 'Yes, I whipped them up the other day . . .'

Sheds can also be calm, meditative places. They tend not to change at the whim of fashion or other fickle influences. They are long-term science projects where the dust and detritus of everyday life is slowly crushed in layers, a bit like the way, given time and pressure, organic matter becomes coal, even eventually diamonds.

A dog can usually sense a good shed. It will walk past, slow down and stop, even wag a tail uncertainly, wishing it could enter. Some finely tuned doggy gland presently unknown to science in its brain senses a certain impossible-to-define lift; a tiny blast of pleasure is available here. A shed like that — approved by dogs — is worth aiming for.

SHEDLOCK

'I've got a shedlock problem — a bit like gridlock in traffic,' says Andrew.

'I've acquired far more material than I'm ever going to use. It's a problem, because I've reached the point where it's too full to work in. Even though nearly all the shelves are on skates and can be moved around, it still takes fifteen minutes to find the stuff I'm looking for and then more time to shift the shelves back. Do that several times a day and you've wasted a lot of time.'

While Andrew may feel he has reached critical shedlock, his shed is still an impressively efficient example of storing electrical and electronic paraphernalia of the late 20th century. Even to get into the heart of his shed involves moving sideways through a small maze of corridors stored from top to bottom with cannibalised parts, all sorted into trays and

shelves. It's a space densely packed with resources and knowledge.

Nevertheless, Andrew is determined to lessen the amount of stuff in the shed, partly because he can see the day when memory loss will make it difficult to function with so much gear. So there are plans to create a bit of open space, maybe enough for a lounge chair, where people can sit down and have a cup of coffee.

Andrew is one of those people who have made the move from the mechanical shed to the electronic shed. His hobby has evolved into his work. He started out in the PMG as a technician where he received comprehensive training, which may explain why he is comfortable working in both the mechanical and the digital worlds. He developed an interest in Ducati motorcycles, which became an activity that filled the shed and then a commercial business, The Bike Factory, separate from home. He later became interested in the early personal computers, making improvements on the now vintage Commodore 64 and Amiga computers. Microcontrollers (microprocessors that are actually small computers for performing a single task) became his special interest and now form the basis for a lot of his work in industrial control processes, robotics and machine tools.

Some of his work involves finding solutions to unusual problems. From working with artists using motors and LEDs to counting oysters and mechatronic heads, life is full of interesting projects.

With that kind of experience, Andrew makes an excellent cannibaliser of electronic parts. He points out that discarded computer monitors, power supplies and printers contain a number of useful electronic parts that can be used to make solar power controllers, battery chargers and all manner of useful electronic components.

Apart from the working electronic parts, there are the glass, plastic and steel elements, which all required huge resources to mine, transform, manufacture and transport. Like many other people with a grasp of just how complex it is to make these things, Andrew is appalled by the waste of resources their thoughtless disposal represents.

To cannibalise is almost a point of honour, even though such scavenging is seen in the wider community as old-fashioned behaviour not suited to our modern times. For most people it's easier to go out and buy something new than to think about how something is made or how it works. That daily flood of junk mail in the letterbox reassures them that purchasing glossy new things best.

Andrew admits that he is as seduced by the packaging and hype as anyone . . . but not totally. He can see the need to reuse the bounty left over by the consumer but doesn't want to be seen as marginalised by taking recycling seriously.

'I see life as being analog or scaled rather than digital or on/off. In all things there is a range, rather than just Holden/Ford or cricket/football etc. Therefore it should be possible for one to be interested in this recycling thing without being seen as totally obsessed. Balance should be possible in life. The trouble is in finding the balance point.'

IT SHOULD BE POSSIBLE FOR ONE TO BE INTERESTED IN THIS RECYCLING THING WITHOUT BEING SEEN AS TOTALLY OBSESSED. BALANCE SHOULD BE POSSIBLE IN LIFE. THE TROUBLE IS IN FINDING THE BALANCE POINT.

ABOVE No point in throwing out perfectly good plugs and leads.
BELOW Components, screws, lights, clips, transistors, resistors, diodes. It's all here.

THE TIME MACHINE

To many people, there's a disturbing darkness to men and their sheds.

There's no doubt that there is a dark side — sheds can be home to suicide, loneliness and isolation. But you can't tell from appearances alone. Andrew's shed might seem at first sight to confirm the negative stereotypes: the skeleton and the coffin are particularly gloomy reminders of the impermanence of life.

In fact, Andrew sees his shed as a positive place. It is important to him because it provides a sort of halfway house between his work as a psychiatric nurse and his home.

'I often come home from work and spend half an hour or an hour here. A space like this allows you to regroup your thoughts. It allows you to go back to your responsibilities refreshed. It's like a workout session

I WORK WITH DISTURBED KIDS AND I SEE THAT THEY MISS THE FABRIC OF HOME LIFE. THEY'RE LOCKED INTO THEIR X-BOXES AND COMPUTER GAMES. WHAT THEY COULD DO IN A SHED JUST ISN'T AVAILABLE TO THEM AND I THINK THEY ARE DEPRIVED OF IMPORTANT THINGS.

or meditation . . . it locks you into a kind of time warp.'

To Andrew it's a non-domestic part of the house which functions like a pressure-relief valve.

'I admit it's a big boy's cubby house. We've had some crazy nights here, carrying on like fools. Sometimes we play cards and chat, or you can just sit down and relax.

'Brand new houses don't have spaces like this in them. That's why I had to build it.'

Even though Andrew only built the shed a couple of years ago and there's not a huge amount of stuff in it (yet), he feels it is a good place to grow the family myths and legends. 'I intend to put up lots of photos of friends and family. And I've started putting things on the wall that I've collected here and there. People talk about that stuff, they tell stories, some true, some tall. Bullshit explodes in here. It's great.

'I want my son to have this sort of experience. I work with disturbed kids and I see that they miss the fabric of home life. They're locked into their X-boxes and computer games. What they could do in a shed just isn't available to them and I think they are deprived of important things.'

Maybe Andrew's shed is a sort of Do-It-Yourself project — not of cupboards or shelves but of self-created family history where layers and layers of shared experiences, stories and things build up to acquire meaning and depth. It's DIY shed geology which, given time, will evolve into something rich.

THE FUTURE OF THE SHED

While the shed has long been seen in overly respectable quarters as some sort of temporary architectural blot on the landscape, sheds and their ilk appear to have a definite future.

This rosy future stems from a number of factors, the most important of which is the need to efficiently gather more of that free water that falls out of the sky now and then. A steel roof structure is still the best way to do that. With the advent of lightweight cold-formed steel structural parts — the parts that were once wood or heavy steel girders — extraordinarily efficient structures are now possible. That efficiency includes the reduced amount of energy required to make the building, the reduced energy required to transport and build with the lighter components.

While it may rust, steel will not warp, rot, split, crack or creep and is not subject to termite attack. Steel is also the most recycled and recyclable material on the planet: a home can be built with the steel from six recycled cars. In Australia we have developed this form of building to new heights. After the tsunami disaster of 2004 several thousand homes were built from lightweight steel structures based on Australian steel building methods.

With the right engineering design, such buildings can even be transported on bicycles and small boats, with the only tools required for construction being a screwdriver and a wrench.

The real breakthrough with this form of design came from the realisation that sheet steel could be easily folded into C-section beams of great strength yet low mass. We lead in this particular building technology partly because there is little historical discrimination here against steel structures as being industrial, vulgar or cheap — which is frequently the case in Europe and America. In fact, if anything, builders are starting to exploit the romance of corrugated iron and its place in Australian history.

So let the sight of sunlight glinting off a corrugated iron roof continue to quicken your heart — not just for the poetry of its shimmering lightness but because its beauty also dwells in its economy, its frugality and now, its environmental benefit.

SAVING THE WORLD

'My favourite thing in the world? It has to be my Rolls-Royce Merlin engine,' says Dave with a cheeky grin.

Dave's wife Suzie is clearly understanding of a man's need for a shed or three — as well as sympathetic to his burning the midnight oil to prepare this impressive engine for insertion into Dave's restoration of *Aggressor*, a famous wooden powerboat of the early 1960s.

'This classic hydroplane was the Australasian champion in its time — undefeated in the 1971–72 season. It's a big brute of an engine for a brute of a boat, which in its day could go an astonishing 170 miles per hour. To think this boat was originally built by someone aged 19 is truly amazing.

'This was the engine that saved the world. In World War Two, it powered aircraft like the Spitfire fighter and the Mustang.

SAVING THE WORLD

'It's a complicated masterpiece. Luckily there are still old mechanics around who know how they worked and I can ask them for help.

'These artefacts — the old race boats and cars — have a social history, a great relevance to our present, past and future. But they are also beautiful things in their own right. To look at this motor — to build something this beautiful which does an incredible task and then to make it successful . . . that would be a great feeling.

'So when you're working on these things, they already have a life or soul of their own. And you working on them, bringing them back to life, means they become a part of you or an extension of yourself. You put so much of yourself into it. It becomes an emotional process, an addiction, maybe even worse than heroin.'

There is something all-consuming about working with very fast cars and boats, their streamlined contours and huge power all pointing in one direction: a few brief seconds or minutes of life on the edge, of barely controlled sound and fury.

It can come crashing down to reality. People get hurt. *Aggressor*'s last race in 1972 left two people fighting for their lives, one permanently disabled. There are more mundane issues, such as the probability that *Aggressor* will use $500 worth of fuel in

a few minutes. But that's just a problem to get around somehow. It's hardly surprising that Dave sees his shed as a lifesaver.

'This shed means access to freedom and happiness, where you can do what you like out of the eye of an increasingly conformist society that frowns on such 'frivolous' activities.

'Without my shed I'm lost. I'm lost without it.'

Photo Courtesy David Jones, Raaf Museum Point Cook

LEFT Dave and his completely restored Merlin powered boat *Aggressor* next to another Merlin powered classic — the Mustang Fighter from World War Two. .

RIGHT Dave's go kart, made from a World War Two aeroplane drop tank. Very stylish.

THE DREAM OF NICKING OFF

The writer John Steinbeck observed of a small wooden boat in a department store that people could not resist the temptation to examine it and knock on its sides to see if it was sound. This reflex may have come from the fact that boats were one of the first man-made tools upon which human life might depend utterly, whether to cross a stretch of otherwise impassable water or to go fishing for a meal.

Perhaps that deep memory is still with us.

According to Robert, fibreglass boats are acceptable only if there is no other choice. But, he maintains, wooden boats have a connection to a history that is almost genetically imbedded in us.

Boats and sheds also have a connected history of very long standing. Robert's shed is the place where he started a successful boat-building

business and to which he now returns twenty years later. He is capitalising on the knowledge built up over that time, trying out customised boat equipment and all the diversions and experiments that sort of fooling around entails.

'It's not really a social shed, so I don't encourage people to come here. Sometimes they come here to hang around and suck your brains, which fortunately doesn't take long, but it takes away from my time staring at the walls wondering if I'm pissing into the wind.'

While this may sound a bit curmudgeonly — a favourite word of his — he is well known as a proselytiser for wooden boats in Australia and around the world. Untold numbers of the nautically innocent have been converted to the cause of the wooden boat by this man's marine

YOU COULD NICK OFF, JUST LIKE THAT. MOST WON'T OF COURSE; FOR THEM THE KNOWING IS FREEDOM ENOUGH.

evangelism. These converts rapidly become addicted to the smell of two-pack epoxy resin as their sheds turn into dusty vortexes of fibreglass, plywood, plans and gummed up power tools.

Robert says it's all in a good cause. 'You come down here [to the shed] and you scratch around, dealing with things you can measure. A hell of a lot of work nowadays deals in things that are intangible or impossible to measure. Here the problems are measurable and tangible and things can be done. The thing is, to get started. It's

that Goethe shit I suppose:

'"Whatever you can do, or dream you can, begin it. Boldness has genius, power, and magic in it. Begin it, now."

'Making wooden boats in your shed is about being part of that pantheon of other blokes who made boats in jungles and caves and huts. Sometimes, late at night I get a sense some of them are taking a look over my shoulder, so I'd better get it right.

'It's also about a dream of freedom. You've got this passport, the wooden boat, which you made, sitting in the shed there, ready to go. You could nick off, just like that. Most won't of course; for them the knowing is freedom enough.

'The shed is a powerful place. You can make freedom there.'
www.nisboats.com

BELOW Although a great deal of epoxy and other synthetic materials are used in most boatbuilding these days, there is still plenty of working in wood required.

RIGHT The boat lurks just outside the shed, constantly whispering 'take me away, take me away . . .'

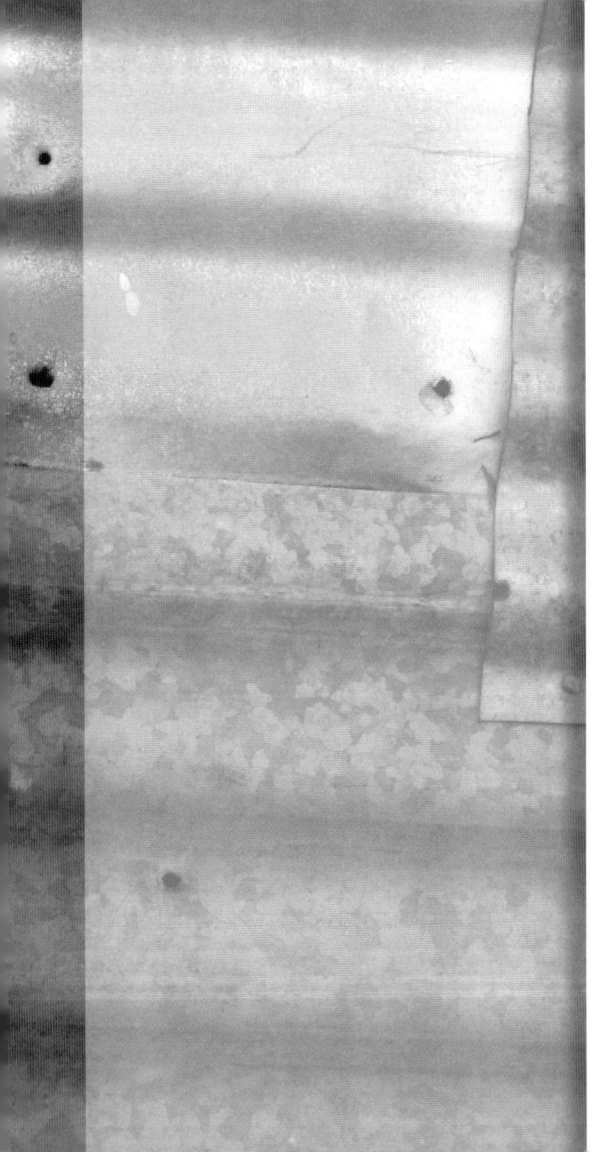

MAX'S MUSEUM

There is almost too much to take in at Max's shed. Apart from the sort of gear you pick up when you've been in the general engineering game since 1948, there are the sculptures. Made from bicycle chains, old sprockets, nuts and bolts, and anything else at hand, Max's whimsical creations have taken over the space. Dozens of candelabras made from bike parts await a match for their candles, and looming over them is an iron stockman (welded from wheel hubs and the like) astride a steel horse with eyes aglow with lightbulbs. Max's creativity with the welder knows no bounds and he still creates the occasional roadside sculpture. 'It keeps me out of the pub,' says Max.

Max doesn't actually own the shed any more — it's owned by a family trust. 'But they let me in to sweep it and look after it.'

The shed is an unofficial local museum filled with exhibits collected in a haphazard way — basically whatever took Max's interest along the way. It has to be said that Max has wide interests.

This is not a museum of the hushed, reverential kind: no white cotton gloves, subdued downlighting or uniformed attendants here — just stuff, Max and lots more stuff, things with stories attached to them. The stories range from episodes of pioneer life to opportunities to crack a corny joke.

Max is not the only person to have this sort of museum. Some people try to make them official, with opening hours and admission fees — the whole box and dice — but that sort of seriousness can crush the pleasure out of the enterprise.

Max has got it about right.

I ALWAYS LIKED A CHALLENGE . . . AND I'VE NEVER BEEN STUMPED YET.

Part of his collection is an impressive array of tractors and some unusual cars. There's a Lightburn Zeta Sports, a stylish Australian-made 500 cc open-topped sports car from 1964. With no doors (you just hop in) and a fibreglass body, it could do 60 miles an hour in reverse. Max used to take the kids to school in it, mostly going forwards, he says.

While Max's shed is a museum of artworks, vehicles and domestic paraphernalia, it is still a place of work should Max need to make or repair anything. Although he has never been formally trained, he has always had a crack at all types of mechanical, electrical and engineering work. In fact, Max says he believes he's held an electrician's licence for longer than anyone in the state.

Like many of the people in this book, Max was the local repair-anything man to whom people brought the tough jobs. As the local farmers were a resourceful bunch, the tough jobs were often very difficult indeed.

'I always knew when they came in and said, "Max, I haven't got time to do this, can you fix it?" that they couldn't do it and it would be a bastard of a job. Still, I always liked a challenge . . . and I've never been stumped yet.'

DO YOU HAVE UNUSUAL 'SHED-MADE' MACHINERY?

If you have custom-made farm machinery or mining equipment, paddock bashers, shooting buggies or, for that matter, anything else cobbled together that works, please contact Mark by emailing:
mark@ibys.org

For more shed culture, and many other things, check out the website of the Australasian Institute of Backyard Studies, where Mark Thomson is the Advanced Research Director.
www.ibys.org

RARE TRADES

MAKING THINGS BY HAND IN THE DIGITAL AGE

by the author of Blokes & Sheds
MARK THOMSON

Dry-stone wallers, coopers, shipwrights — the people in this book make things with their hands. They take raw materials and transform them, with considerable skill, into objects that are always useful and often beautiful. What can we learn from these experienced artisans in an age when our hands are primarily used to press the keys on our computers and the buttons on our telephones? Is the art of making all but lost, except to just a handful? Is that why our local hardware shops cover five hectares — are we looking for an excuse to grasp a brush or swing a hammer?

Rare Trades offers a fascinating insight into the tools, methods and philosophies of skilled artisans. Author and photographer Mark Thomson (*Blokes & Sheds*, *Stories from the Shed*, and *Blokes & Barbies*) travels with camera and notebook through the backyards and backroads of Australia to find people who still make things the old way. On his journey he discovers that patience, integrity and acute observation are as important to these people as being 'good with their hands', and proves that the human urge to make things by hand is deep and irresistible.